Preface

The publication of BS 7671 and its predecessors, the 15th and 16th Editions of the IEE Wiring Regulations, led to a number of guides and handbooks published by organisations involved in the electrical contracting industry. These included the publication, by the Institution of Electrical Engineers, of an On-site Guide and a number of Guidance Notes as well as several books by independent authors and a considerable number of articles and papers in the technical press. It also led to numerous instructional courses, seminars and conferences.

It was thought that there was little else one could write about concerning the Wiring Regulations, but after talking to a number of engineers in the electrical installation contracting industry, Brian Jenkins gained the strong impression that there was one need which had not really been satisfied. The need was for a book that made considerable use of worked examples with the absolute minimum discussion of the associated theoretical aspects. In other words, a book which used such examples to show how one carried out the calculations involved in circuit design for compliance with BS 7671.

Whilst Brian designed the book to be primarily of interest and help to those in the smaller companies in the electrical installation contracting industry, we believe the student and the plant engineer will also find it of interest.

BS 7671 offers certain options. For example, when calculating voltage drop either an approximate method or a more accurate one can be used and we have attempted to show where the latter could be used to advantage. This, we believe, will make the book of interest to a wider circle.

BS 7671, like the 16th Edition, does not refer to 'touch voltages' as such, these being the 'voltages between simultaneously accessible exposed and extraneous conductive parts' mentioned in the key regulation, Regulation 413-02-04, concerning automatic disconnection of supply as the protective measure against indirect contact.

It has long been Brian's opinion that a fuller understanding of the touch voltage concept would assist many in the electrical contracting industry to more fully understand the requirements for automatic disconnection. For this reason we hope that the Appendix will prove to be of interest.

Since the first edition of this book there have been a number of amendments to the Requirement for Electrical Installations. Some of the changes introduced by the amendments affect the examples given in this book. The most important changes have been the change to the nominal voltage from 240/415 V to 230/400 V and the change to the assumed temperature of conductors under fault conditions. Changes to other British Standards have necessitated changes in some examples, e.g. the withdrawal of BS 3871. This third edition is intended to keep *Electrical Installation Calculations* up to date with the latest version of BS 7671.

There is one final point which needs to be made in this Preface. Examination of some of the answers may suggest to the reader that there is a high intrinsic degree of accuracy in installation design calculations. This obviously cannot be true because, for example, estimated circuit lengths will be rather approximate.

Many of the answers have been given to a greater number of significant figures than is necessary in practice merely to assist the reader should he, or she, wish to check through the examples.

Mark Coates

Electrical Installation Calculations

For compliance with BS 7671: 2001
(The Wiring Regulations)

Third I

B.D. J

M. Co

Publishe

Represent
engineeri

Blackwell
Science

Blackwell Science Ltd, a Blackwell Publishing company
Editorial Offices:
Blackwell Science Ltd, 9600 Garsington Road, Oxford OX4 2DQ, UK
Tel: +44 (0) 1865 776868
Blackwell Publishing Inc., 350 Main Street, Malden, MA 02148-5020, USA
Tel: +1 781 388 8250
Blackwell Science Asia Pty, 550 Swanston Street, Carlton, Victoria 3053, Australia
Tel: +61 (0)3 8359 1011

First edition published 1991
Reprinted 1991, 1992 (twice), 1993, 1994, 1996
Second edition published 1998
Third edition published 2003
Reprinted 2006

ISBN-10: 0-632-06485-4
ISBN-13: 978-0-632-06485-4

Library of Congress Cataloging-in-Publication Data
Jenkins, Brian D. (Brian David)
Electrical installation calculations : for compliance with BS 7671:2001 (the wiring regulations) / B.D. Jenkins — 3rd ed.
p. cm.
"Published on behalf of ECA."
ISBN 0-632-06485-4 (alk. paper)
1. Electric wiring—Mathematics. I. Coates, M. (Mark) II. Electrical Contractors' Association (Great Britain) III. Title.

TK3211 .J44 2002
621.319'24'0151—dc21

2002035641

A catalogue record for this title is available from the British Library

Set in Times
by Aarontype Ltd, Bristol
Printed and bound in Great Britain
by Marston Book Services, Oxford

For further information on Blackwell Publishing, visit our website:
www.blackwellpublishing.com

Contents

Preface *v*
Acknowledgements *vii*
Symbols *ix*
Definitions *xi*

1 Calculation of the Cross-sectional Areas of Circuit Live Conductors **1**
General circuits 3
Circuits in thermally insulating walls 5
Circuits totally surrounded by thermally insulating material 6
Circuits in varying external influences and installation conditions 6
Circuits in ventilated trenches 9
Circuits using mineral-insulated cables 10
Circuits on perforated metal cable trays 11
Circuits in enclosed trenches 12
Grouped circuits not liable to simultaneous overload 13
Circuits in low ambient temperatures 19
Grouped ring circuits 21
Motor circuits subject to frequent stopping and starting 22
Circuits for star-delta starting of motors 24
Change of parameters of already installed circuits 25
Admixtures of cable sizes in enclosures 28
Grouping of cables having different insulation 34

2 Calculation of Voltage Drop Under Normal Load Conditions **35**
The simple approach 35
The more accurate approach taking account of conductor operating temperature 39
The more accurate approach taking account of load power factor 51
The more accurate approach taking account of both conductor operating temperature and load power factor 53
Voltage drop in ring circuits 55
Voltage drop in ELV circuits 57

3 Calculation of Earth Fault Loop Impedance **60**
The simple approach 64
The more accurate approach taking account of conductor temperature 68
Calculations taking account of transformer impedance 74
Calculations concerning circuits fed from sub-distribution boards 76
Calculations where conduit or trunking is used as the protective conductor 81
Calculations where cable armouring is used as the protective conductor 88

4 Calculations Concerning Protective Conductor Cross-sectional Areas **93**
Calculations when the protective device is a fuse 96
Calculations when the protective device is an mcb 102
Calculations when the protective device is an rccb 108

5 Calculations Related to Short Circuit Conditions **116**
A.C. single-phase circuits 117
The more rigorous method for A.C. single-phase circuits 124
A.C. three-phase circuits 130

6 Combined Examples **141**

Appendix The Touch Voltage Concept 163
Index 175

Acknowledgements

Brian Jenkins acknowledges the initial encouragement and subsequent assistance given by M.J. Dyer when he was Director of Technical Services of the Electrical Contractors' Association and by C.P. Webber BTech, CEng, MIEE, the present Head of Technical Services of that Association.

He also wishes to acknowledge the considerable assistance given by a number of friends who kindly agreed to read his drafts and who offered useful suggestions. In this respect he particularly wishes to thank:

F.W. Price, CEng, MIEE
J. Rickwood, BSc (Eng), CEng, FIEE
G. Stokes, BSc, CEng, MIEE, FCIBSE, MISOH
J.F. Wilson, MBE, AMIEE

Finally, thanks are due to the Institution of Electrical Engineers for its permission to reproduce a number of the definitions from BS 7671 and the International Electro-technical Commission for their permission to reproduce the touch voltage curves shown in the Appendix. These curves are in an IEC Technical Report IEC 1200-413.

In the compilation of the second and third edition Mark Coates wishes to acknowledge the help of P.J. Buckle and D. Locke, both of the Electrical Contractors' Association, for their assistance as advisors and critics.

Symbols

The symbols used in this book are generally aligned with those used in BS 7671 together with some additional symbols which have been found necessary.

Symbols used infrequently are defined where they occur in the text.

C_a correction factor for ambient temperature

C_d correction factor for type of overcurrent protective device
$C_d = 1$ for HBC fuses and mcbs
$C_d = 0.725$ for semi-enclosed fuses

C_g correction factor for grouping

C_i correction factor for conductors embedded in thermal insulation

C_r correction factor for grouping of ring circuits

l circuit route length, m

I_b design current of circuit, A

$I_{\Delta n}$ rated residual operating current of an rcd, mA or A

I_{ef} earth fault current, A

I_n nominal current of protective device, A

I_{sc} short circuit current, A

I_t required tabulated current-carrying capacity, A

I_{ta} actual tabulated current-carrying capacity, A

I_x current used as a basis for calculating the required current-carrying capacity of the live conductors, A

I_z effective current-carrying capacity, A

S conductor cross-sectional area, mm^2

t_a actual or expected ambient temperature, °C

t_o maximum permitted conductor temperature under overload conditions, °C

t_p	maximum permitted normal operating conductor temperature, °C
t_r	reference ambient temperature, °C – (t_r in BS 7671 is 30°C)
t_1	actual conductor operating temperature, °C
U_n	nominal voltage, V
U_o	nominal voltage to Earth, V
U_p	nominal phase voltage, V
Z_1	impedance of live conductor, ohms $=\sqrt{(R_1{}^2+X_1{}^2)}$ where R_1 is its resistance component and X_1 is its reactance component
Z_2	impedance of protective conductor, ohms $=\sqrt{(R_2{}^2+X_2{}^2)}$ where R_2 is its resistance component and X_2 is its reactance component
Z_E	that part of the earth fault loop impedance which is external to the installation, ohms
Z_{pn}	phase to neutral impedance, ohms
Z_s	earth fault loop impedance, ohms

Definitions

The following definitions are of terms which appear in this book and have been aligned, generally without modification, with the definitions in BS 7671: 1992.

Ambient temperature
The temperature of the air or other medium where the equipment is to be used.

Bonding conductor
A protective conductor providing equipotential bonding.

Bunched
Cables are said to be bunched when two or more are contained within a single conduit, duct, ducting, or trunking or, if not enclosed, are not separated from each other by a specified distance.

Circuit protective conductor (cpc)
A protective conductor connecting exposed conductive parts of equipment to the main earthing terminal.

Current-carrying capacity of a conductor
The maximum current which can be carried by a conductor under specified conditions without its steady state temperature exceeding a specified value.

Design current (of a circuit)
The magnitude of the current (rms value for a.c.) to be carried by the circuit in normal service.

Direct contact
Contact of persons or livestock with live parts.

Distribution circuit
A Band II circuit connecting the origin of the installation to:

- an item of switchgear, or
- an item of controlgear, or
- a distribution board

to which one more more final circuits or items of current-using equipment are connected (see also definition of final circuit). A distribution circuit may also connect the origin of an installation to an outlying building or separate installation, when it is sometimes called a sub-mains.

Earth fault current
A fault current which flows to earth.

Earth fault loop impedance
The impedance of the earth fault current loop starting and ending at the point of earth fault. This impedance is denoted by Z_S.

The earth fault loop comprises the following, starting at the point of fault:

- the circuit protective conductor
- the consumer's earthing terminal and earthing conductor
- for TN systems, the metallic return path
- for TT and IT systems, the earth return path
- the path through the earthed neutral point of the supply transformer and the transformer winding
- the phase conductor from the transformer supply to the point of fault.

Earth leakage current
A current which flows to earth, or to extraneous-conductive-parts, in a circuit which is electrically sound. This current may have a capacitive component including that resulting from the deliberate use of capacitors.

Earthing
Connection of the exposed-conductive-parts of an installation to the main earthing terminal of that installation.

Earthing conductor
A protective conductor connecting the main earthing terminal of an installation to an earth electrode or to other means of earthing.

Equipotential bonding
Electrical connection maintaining various exposed-conductive-parts and extraneous-conductive-parts at substantially the same potential.

Exposed-conductive-part
A conductive part of equipment which can be touched and which is not a live part but which may become live under fault conditions.

External influence
Any influence external to an electrical installation which affects the design and safe operation of that installation.

Extraneous-conductive-part
A conductive part liable to introduce a potential, generally earth potential, and not forming part of the electrical installation.

Fault current
A current resulting from an insulation failure or the bridging of insulation.

Final circuit
A circuit connected directly to current-using equipment, or to a socket-outlet or socket-outlets, or other outlet points for the connection of such equipment.

Indirect contact
Contact of persons or livestock with exposed-conductive-parts which have become live under fault conditions.

Live part
A conductor or conductive part intended to be energised in normal use, including a neutral conductor but, by convention, not a PEN conductor.

Main earthing terminal
The terminal or bar provided for the connection of protective conductors, including equipotential bonding conductors, and conductors for functional earthing if any, to the means of earthing.

Origin of an installation
The position at which electrical energy is delivered to an electrical installation.

Overcurrent
A current exceeding the rated value. For conductors the rated value is the current-carrying capacity.

Overload current
An overcurrent occurring in a circuit which is electrically sound.

PEN conductor
A conductor combining the functions of both protective conductor and neutral conductor.

Protective conductor
A conductor used for some measures of protection against electric shock and intended for connecting together any of the following parts:

- exposed-conductive-parts
- extraneous-conductive-parts
- the main earthing terminal
- earth electrode(s)
- the earthed point of the source, or an artificial neutral.

Residual current
The algebraic sum of the currents flowing in the live conductors of a circuit at a point in the electrical installation.

Residual current device
A mechanical switching device or association of devices intended to cause the opening of the contacts when the residual current attains a given value under specified conditions.

Residual operating current
Residual current which causes the residual current device to operate under specified conditions.

Ring final circuit
A final circuit arranged in the form of a ring and connected to a single point of supply.

Short circuit current
An overcurrent resulting from a fault of negligible impedance between live conductors having a difference in potential under normal operating conditions.

System
An electrical system consisting of a single source of electrical energy and an installation. For certain purposes (of the Wiring Regulations), types of system are identified as follows, depending upon the relationship of the source, and of exposed-conductive-parts the installation, to Earth:

- *TN system*, a system having one or more points of the source of energy directly earthed, the exposed-conductive-parts of the installation being connected to that point by protective conductors.
- *TN–C system*, in which neutral and protective functions are combined in a single conductor throughout the system.
- *TN–S system*, having separate neutral and protective conductors throughout the system.
- *TN–C–S system*, in which neutral and protective functions are combined in a single conductor in part of the system.
- *TT system*, a system having one point of the source of energy directly earthed, the exposed-conductive-parts of the installation being connected to earth electrodes electrically independent of the earth electrodes of the source.
- *IT system*, a system having no direct connection between live parts and earth, the exposed-conductive-parts of the electrical installation being earthed.

Voltage, nominal
Voltage by which an installation (or part of an installation) is designated. The following ranges of nominal voltage (rms values for a.c.) are defined:

- *Extra-low*. Normally not exceeding 50 V a.c. or 120 V ripple free d.c., whether between conductors or to earth.
- *Low*. Normally exceeding extra-low voltage but not exceeding 1000 V a.c. or 1500 V d.c. between conductors, or 600 V a.c. or 900 V d.c. between conductors and earth.

The actual voltage of the installation may differ from the nominal value by a quantity within normal tolerances.

Chapter 1

Calculation of the Cross-sectional Areas of Circuit Live Conductors

The first stage in designing an installation after having carried out the assessment of general characteristics demanded in Part 3 of BS 7671 is the choice of the type of cable and the method of installation of that cable for each circuit. In some cases these choices are closely interrelated, e.g. non-sheathed cables are required to be enclosed in conduit, duct, ducting or trunking (Regulation 521–07–03).

Where there are a number of options open to the installation designer from purely technical considerations, the final choice will depend on commercial aspects or the designer's (or client's) personal preferences. Here it is assumed that the designer, after having taken into account the relevant external influences to which the circuit concerned is expected to be subjected, has already decided on the type of cable and the installation method to use. The appropriate table of current-carrying capacity, in Appendix 4 of BS 7671, and the appropriate column within that table are therefore known.

To determine the minimum conductor cross-sectional area of the live conductors of a particular circuit that can be tolerated the designer *must*:

(a) Establish what is the expected ambient temperature (t_a°C). This gives the relevant value of C_a. Note that more than one value of t_a°C may be encountered in some installations. Where there is more than one value the designer may opt to base all his calculations on the highest value or, alternatively, base his calculations for a particular part of the installation on the value of t_a°C pertinent to that part.

(b) Decide whether the circuit is to be run singly or be bunched or grouped with other circuits and, if the latter, how many other circuits. The decision taken gives the relevant value of C_g.

(c) Decide whether the circuit is likely to be totally surrounded by thermally insulating material (Regulation 523–04–01). If this is to be the case, C_i is taken to be 0.5.

(d) Determine the design current (I_b) of the circuit, taking into account diversity where appropriate (Regulation 311–01–01), and any special characteristics of the load, e.g. motors subject to frequent stopping and starting (Regulation 552–01–01).

(e) Choose the type and nominal current rating (I_n) of the associated overcurrent protective device. For *all* cases I_n must be equal to or greater than I_b. Remember that overcurrent protective devices must comply with Chapter 43 of BS 7671 as regards their breaking capacity, but for the present let it be assumed the chosen devices do so comply.

(f) Establish whether it is intended the overcurrent protective device is to give:
   (i) overload protection only, or
   (ii) short circuit protection only, or
   (iii) overload *and* short circuit protection.

The intended function of the overcurrent protective device not only determines whether I_b or I_n is used as the basis for calculating the minimum cross-sectional area of the live conductors but also influences the value of C_a that is to be used in the calculations.

(g) Establish the maximum voltage drop that can be tolerated.
(h) Estimate the route length of the circuit.

It cannot be emphasised too strongly that unless all the foregoing items are available it is not possible to design any circuit.

The general method for the determination of the minimum conductor cross-sectional areas that can be tolerated now described does *not* apply to cables installed in *enclosed* trenches. These are considered later in this chapter. The general method is as follows:

First calculate the current I_t where:

$$I_t = I_x \times \frac{1}{C_a} \times \frac{1}{C_g} \times \frac{1}{C_i} \times \frac{1}{C_d} \text{ A}$$

Table 1.1 indicates when $I_x = I_b$ and when $I_x = I_n$. That table also indicates from which Tables in Appendix 4 of BS 7671 the appropriate values of C_a and C_g are found and it gives the values of C_d to use.

Table 1.1 Selection of current and of correction factors.

Overcurrent protective device	Protection provided	For I_x use	Obtain C_a from	Obtain C_g from	C_d
Semi-enclosed fuses to BS 3036	Overload with or without short circuit	I_n	Table 4C2	Table 4B1	0.725
	Short circuit only	I_b	Table 4C1	Table 4B1	1
HBC fuses to BS 88 Pt 2 or Pt 6 or BS 1361 or mcbs	Overload with or without short circuit	I_n	Table 4C1	Table 4B1	1
	Short circuit only	I_b	Table 4C1	Table 4B1	1

C_i (when applicable) is, in the absence of more precise information, taken to be 0.5 and the calculation is then based on the tabulated current-carrying capacity for Reference Method 1, i.e. for cables clipped direct to a surface and open. It should be noted that even when the cable concerned is installed in thermal insulation for comparatively short lengths (up to 400 mm) Regulation 523–04–01 specifies derating factors varying from 0.89 to 0.55.

Having calculated I_t, all that remains to do is to inspect the appropriate table of current-carrying capacity in Appendix 4 and the appropriate column in that table in order to find that conductor cross-sectional area having an *actual* tabulated current-carrying capacity (I_{ta}) equal to or greater than the calculated I_t.

Note that in the following examples the circuit lengths are not given and therefore the voltage drops are not calculable. The examples are concerned solely with the determination of conductor cross-sectional areas for compliance with the requirements in BS 7671 regarding the thermal capability of cables under normal load conditions and, where appropriate, under overload conditions.

It should be remembered that Regulation 434–03–02 allows the designer to assume that, if the overcurrent protective device is intended to provide both overload and short circuit protection, there is no need to carry out further calculation to verify compliance with the requirement (given in Regulation 434–03–03) regarding the latter. That assumption has been made in the following examples. However, when the overcurrent protective device is intended to provide short circuit protection only (e.g. in motor circuits) it is essential that the calculations described in Chapter 5 are made.

Overcurrent protective devices are also frequently intended to provide automatic disconnection of the supply in the event of an earth fault (i.e. to provide protection against indirect contact) and the calculations that are then necessary are described in Chapter 3.

In practice, particularly with single-phase circuits which are not bunched, it will be found that voltage drop under normal load conditions determines the cross-sectional area of the circuit live conductors that can be used and it is for this reason that Chapter 2 deals with voltage drop calculations as these are sensibly the next stage in circuit design.

GENERAL CIRCUITS

Example 1.1

A single-phase circuit is to be wired in 70°C pvc-insulated and sheathed single-core cables to BS 6004 having copper conductors. The cables are to be installed in free air, horizontal, flat spaced on cable supports (Reference Method 12).

If $I_b = 135$ A, $t_a = 50$°C and overload and short circuit protection is to be provided by a BS 88 'gG' fuse (BS EN 60269), what is the minimum current rating for that fuse and the minimum cross-sectional area of the cable conductors that can be used?

Answer

Because $I_n \geqslant I_b$, select I_n from the standard ratings of BS 88 Pt 2.

Therefore $I_n = 160$ A.

From Table 4C1, $C_a = 0.71$.

As there is no grouping, $C_g = 1$.

Also $C_i = 1$ and $C_d = 1$.

Thus:

$$I_t = 160 \times \frac{1}{0.71} \text{ A} = 225.4 \text{ A}$$

From Table 4D1A Column 10 it is found that 50 mm^2 is inadequate because I_{ta} would be only 219 A. The minimum conductor cross-sectional area that can be used is 70 mm^2 having $I_{ta} = 281$ A.

Example 1.2

A d.c. circuit has a design current of 28 A. It is to be wired in a two-core cable to BS 6883 having 85°C rubber insulation and copper conductors. It is to be installed in trunking with five other similar circuits.

If $t_a = 40°C$ and the circuit is to be protected by a 45 A HBC fuse to BS 1361 against short circuit faults only, what is the minimum conductor cross-sectional area that can be used?

Answer

From Table 4B1, $C_g = 0.57$ (there being a total of 6 circuits).

From Table 4C1, $C_a = 0.90$.

Also $C_i = 1$ and $C_d = 1$.

Thus:

$$I_t = 28 \times \frac{1}{0.90} \times \frac{1}{0.57} \text{ A} = 54.6 \text{ A}$$

From Table 4F2A Column 2 the minimum conductor cross-sectional area that can be used is found to be 10 mm^2 having $I_{ta} = 66$ A.

Note that in this example the I_b of 28 A is used to determine I_t because only short circuit protection is intended and that it is still necessary to check that the 10 mm^2 conductors will comply with Regulation 434 03 03, using the procedure described in Chapter 5.

Example 1.3

A single-phase circuit has $I_b = 17$ A and is to be wired in flat two-core (with cpc) 70°C pvc-insulated and sheathed cable to BS 6004 having copper conductors, grouped with four other similar cables, all clipped direct.

If $t_a = 45°C$ and the circuit is to be protected against both overload and short circuit by a semi-enclosed fuse to BS 3036 what should be the nominal current rating of that fuse and the minimum cross-sectional area of cable conductor?

Answer

Because $I_n \geqslant I_b$ select from the standard ratings for BS 3036 fuses $I_n = 20\,A$.

From Table 4B1, $C_g = 0.60$ (there being a total of 5 circuits).

From Table 4C2, $C_a = 0.91$.

Also $C_i = 1$ and $C_d = 0.725$.

Thus:

$$I_t = 20 \times \frac{1}{0.91} \times \frac{1}{0.60} \times \frac{1}{0.725}\,A = 50.5\,A$$

From Table 4D5A Column 4 it is found that the minimum conductor cross-sectional area that can be used is $10\,mm^2$ having $I_{ta} = 64\,A$.

CIRCUITS IN THERMALLY INSULATING WALLS

Example 1.4

Five similar three-phase circuits each having $I_b = 30\,A$ are to be wired in single-core 70°C pvc-insulated non-sheathed cables to BS 6004 having copper conductors. The circuits are installed in trunking in a thermally insulating wall, the trunking being in contact with a thermally conductive surface on one side (Reference Method 4).

If the circuits are protected against both overload and short circuit by 32 A BS 88 'gG' fuses and $t_a = 45°C$, what is the minimum conductor cross-sectional area that can be used?

Answer

From Table 4B1, $C_g = 0.60$.

From Table 4C1, $C_a = 0.79$.

Also $C_i = 1$ and $C_d = 1$.

Thus:

$$I_t = 32 \times \frac{1}{0.79} \times \frac{1}{0.60}\,A = 67.5\,A$$

From Table 4D1A Column 3 it is found that the minimum conductor cross-sectional area that can be used is $25\,mm^2$ having $I_{ta} = 73\,A$.

CIRCUITS TOTALLY SURROUNDED BY THERMALLY INSULATING MATERIAL

Example 1.5

Six similar single-phase circuits each having $I_b = 8$ A are to be wired in single-core 70°C pvc-insulated non-sheathed cables to BS 6004 having copper conductors. The cables are enclosed in conduit totally surrounded by thermally insulating material.

If the ambient temperature is expected to be 45°C and each circuit is protected by a 10 A miniature circuit breaker against both overload and short circuit, what is the minimum cross-sectional area of conductor that can be used?

Answer

From Table 4B1, $C_g = 0.57$.

From Table 4C1, $C_a = 0.79$.

The relevant table of current-carrying capacity is Table 4D1A. Column 2 must *not* be used because it relates to Reference Method 4 where the conduit enclosing the cable(s) is in contact with a thermally conductive surface on one side. In this example the conduit is totally surrounded by thermally insulating material: Regulation 523–04–01 indicates that in such cases the factor C_i (=0.5) has to be related to the current-carrying capacities for cables clipped direct (Reference Method 1). Thus:

$$I_t = 10 \times \frac{1}{0.79} \times \frac{1}{0.57} \times \frac{1}{0.5}\ \text{A} = 44.4\ \text{A}$$

From Column 6 of Table 4D1A it is found that the minimum conductor cross-sectional area that can be used is 6 mm^2 having $I_{ta} = 47$ A.

Note that exactly the same procedure must be used if the cable(s) concerned are not in conduit but still totally surrounded by thermally insulating material. Regulation 523–04–01, as already indicated, gives derating or correction factors for cables installed in thermal insulation for comparatively short lengths, i.e. up to 0.4 m.

CIRCUITS IN VARYING EXTERNAL INFLUENCES AND INSTALLATION CONDITIONS

In practice, the problem is frequently encountered that a circuit is so run that the relevant values of C_a and/or C_g are not constant or applicable over the whole circuit length. In such cases, *provided that the Reference Method does remain the same*, the quickest way of determining the minimum conductor cross-sectional area that can be used is to calculate the product C_aC_g for each section of the circuit and then to use the *lowest* such product in the equation for determining I_t.

Example 1.6

A single-phase circuit is to be wired in single-core 70°C pvc-insulated non-sheathed cables to BS 6004 having copper conductors and protected against both overload and short circuit by a 25 A BS 88 'gG' fuse. For approximately the first third of its route it is run in trunking with six other similar circuits in an ambient temperature of 30°C. For the remaining two-thirds of its route it is run in conduit on a wall but with no other circuits and where the ambient temperature is 50°C.

Determine the minimum conductor cross-sectional area that can be used.

Answer

The method of installation is Reference Method 3 throughout, so proceed as follows.

For the first third of the route

$C_a = 1$ and, from Table 4B1, $C_g = 0.54$.

Thus: $C_aC_g = 0.54$

For the remaining two-thirds of the route

From Table 4C1, $C_a = 0.71$ but $C_g = 1$

Thus: $C_aC_g = 0.71$

The more onerous section of the route is therefore the first third, i.e. in the trunking and

$$I_t = \frac{25}{0.54}\ \text{A} = 46.3\ \text{A}$$

From Table 4D1A Column 4 it is found that the minimum conductor cross-sectional area that can be used is 10 mm^2 having $I_{ta} = 57$ A. The designer has two options. Either the whole circuit is run in 10 mm^2 or a check can be made to find out if it would be possible to reduce the cross-sectional area over that part of the circuit run singly in conduit.

In the present case for that part:

$$I_t = \frac{25}{0.71}\ \text{A} = 35.2\ \text{A}$$

Again from Table 4D1A Column 4 it would be found that a conductor cross-sectional area of 6 mm^2 could be used.

Whether the designer opts to take advantage of this is a matter of personal choice.

Now consider the case where the Reference Method is *not* the same throughout. Then the procedure to use is to treat each section separately and using the appropriate values of the relevant correction factors calculate the required I_t.

Example 1.7

A three-phase circuit having $I_b = 35\,A$ is to be wired in multicore non-armoured 70°C pvc-insulated and sheathed cables having copper conductors and is protected against both overload and short circuit by a TP&N 40 A miniature circuit breaker. For approximately three-quarters of its length it is run in trunking with four other similar circuits in an expected ambient temperature of 25°C. For the remaining part of its route it is run singly clipped direct to a wall but where the ambient temperature is 45°C.

Determine the minimum conductor cross-sectional area that can be used.

Answer

When grouped in trunking:

From Table 4B1, $C_g = 0.6$.

From Table 4C1, $C_a = 1.03$.

Thus:

$$I_t = 40 \times \frac{1}{1.03} \times \frac{1}{0.60}\,A = 64.7\,A$$

From Table 4D2A Column 5 the minimum conductor cross-sectional area that can be used is found to be $25\,mm^2$ having $I_{ta} = 80\,A$.

When run singly, clipped direct:

From Table 4C1, $C_a = 0.79$.

Also $C_g = 1$.

Thus:

$$I_t = 40 \times \frac{1}{0.79}\,A = 50.6\,A$$

From Table 4D2A Column 7 the minimum conductor cross-sectional area that can be used is found to be $10\,mm^2$ having $I_{ta} = 57\,A$.

Whether, for the circuit concerned, the designer opts to reduce the cable cross-sectional area where it exits from the trunking, is a matter of personal choice.

CIRCUITS IN VENTILATED TRENCHES

Example 1.8

Four similar single-phase circuits are to be wired in single-core non-armoured 70°C pvc-insulated and sheathed cables having aluminium conductors. These cables are to be supported on the walls of a ventilated trench (Installation Method 17) vertically spaced in accordance with Reference Method 12.

If, for each circuit, $I_b = 175\,A$ and each circuit is to be protected by a 200 A BS 88 'gG' fuse against short circuit current only, what is the minimum conductor cross-sectional area that can be used, t_a being 50°C?

Answer

Inspection of Table 4B1 shows that no grouping factor (C_g) is given for Reference Method 12. Provided the cable spacing indicated in Table 4A for that Method is used, $C_g = 1$.

From Table 4C1, $C_a = 0.71$.

Thus:

$$I_t = 175 \times \frac{1}{0.71}\,A = 246.5\,A$$

From Table 4K1A Column 11 the minimum conductor cross-sectional area that can be used is found to be $120\,mm^2$ having $I_{ta} = 273\,A$.

Example 1.9

Seven similar three-phase circuits are to be wired in multicore non-armoured 70°C pvc-insulated and sheathed cables having copper conductors. These cables are to be supported on the walls of a ventilated trench (Installation Method 17) spaced in accordance with Reference Method 13.

If, for each circuit, $I_b = 55\,A$ and each circuit is to be protected by 63 A BS 88 'gG' fuses against both overload and short circuit, what is the minimum conductor cross-sectional area that can be used, t_a being 35°C?

Answer

Inspection of Table 4B1 shows that no grouping factor (C_g) is given for Reference Method 13. Provided the cable spacing indicated in Table 4A for that Method is used, $C_g = 1$.

From Table 4C1, $C_a = 0.94$.

Thus:

$$I_t = 63 \times \frac{1}{0.94} \text{ A} = 67 \text{ A}$$

From Table 4D2A Column 9 the minimum conductor cross-sectional area that can be used is found to be 16 mm^2 having I_{ta} = 80 A.

It is to be noted that in Table 4A the only spacing requirement given for Reference Method 13 is the distance between the cable surface and the trench wall. No indication is given as regards the minimum distance between cables. It is suggested that the vertical spacing should be at least three times that given for single-core cables in Reference Method 12 but the horizontal spacing can be the same as that given for single-core cables. Because of the restricted air flow in trenches, even in ventilated trenches, the number of horizontal layers of cables should be limited. Of course, wherever possible, the spacing between cables should be as large as the trench dimensions allow. It should be noted that, for large single-core cables, an increase in the spacing will increase the voltage drop.

CIRCUITS USING MINERAL-INSULATED CABLES

The next two examples illustrate two important points concerning mineral-insulated cables.

Example 1.10

A single-phase circuit is run in light duty mineral-insulated cable, clipped direct and having an overall covering of pvc. The circuit is protected by a 25 A semi-enclosed fuse to BS 3036 against both overload and short circuit.

If t_a = 45°C, what is the minimum conductor cross-sectional area that can be used?

Answer

From Table 4C2, C_a = 0.89.

Thus:

$$I_t = 25 \times \frac{1}{0.89} \text{ A} = 28.1 \text{ A}$$

From Table 4J1A Column 2 the minimum conductor cross-sectional area that can be used is found to be 2.5 mm^2 having I_{ta} = 31 A.

It is important to note that although protection is being provided by a semi-enclosed fuse it is *not* necessary to use the 0.725 factor normally associated with such fuses in order to determine I_t.

A further point to note is that had the cable been bare and exposed to touch, Note 2 to Table 4J1A indicates that the tabulated values have to be multiplied by 0.9 and I_{ta} would then have been 27.9 A. It would then have been necessary to use 4 mm^2 conductors.

Example 1.11

Six similar single-phase circuits are run in heavy duty single core mineral-insulated cables, bare and not exposed to touch, bunched and clipped direct.

If each circuit is protected by a 50 A BS 88 'gG' fuse against both overload and short circuit and $t_a = 35°C$, what is the minimum conductor cross-sectional area that can be used?

Answer

From Table 4C1, for 105°C sheath temperature, $C_a = 0.96$.

Thus:

$$I_t = 50 \times \frac{1}{0.96}\ A = 52.1\ A$$

From Table 4J2A Column 2 the minimum conductor cross-sectional area that can be used is found to be 4 mm^2 having $I_{ta} = 55$ A.

Note that as indicated in Note 2 to Table 4J2A it is *not* necessary to apply a grouping factor. For mineral-insulated cables exposed to touch or having pvc covering it is, however, necessary to apply grouping factors, as given in Table 4B1 or, if on perforated trays, Table 4B2.

CIRCUITS ON PERFORATED METAL CABLE TRAYS

Example 1.12

Six similar three-phase circuits are to be run in multicore non-armoured cables having 85°C rubber insulation and copper conductors, installed as a single layer on a perforated metal cable tray.

If each circuit is protected against both overload and short circuit by a 50 A mcb and $t_a = 60°C$, what is the minimum conductor cross-sectional area that can be used?

Answer

Examination of Table 4B1 giving values of C_g shows that the values are given for cables touching and for cables spaced (where 'spaced' is defined as a clearance between adjacent surfaces of at least one cable diameter).

Not enough information has been given in the example and the assumption has to be made that the cables are touching. C_g is therefore 0.74.

From Table 4C1, $C_a = 0.67$.

Thus:

$$I_t = 50 \times \frac{1}{0.74} \times \frac{1}{0.67} \text{ A} = 100.8 \text{ A}$$

From Table 4F2A Column 7 it is found that the minimum conductor cross-sectional area that can be used is $25\,\text{mm}^2$ having $I_{ta} = 123$ A.

Where circuits on trays use a variety of conductor sizes in multiple layers ERA Publications 69–30 Parts 6 and 7 give a method of calculation of those sizes.

CIRCUITS IN ENCLOSED TRENCHES

As indicated earlier the procedure adopted for the foregoing examples cannot be used for cables installed in *enclosed* trenches.

This is because the grouping factors given in Table 4B1 do not apply to this method of installation and one has to use the correction factors (denoted here by C_e) given in Table 4B3 which are related to conductor cross-sectional areas. An iterative procedure has therefore to be used as illustrated by Example 1.13.

Example 1.13

A three-phase circuit having $I_b = 55$ A is installed in an enclosed trench with 5 other similar circuits. The circuits are to be run in multi-core armoured 70°C pvc-insulated cables to BS 6346 having copper conductors and each is to be protected against both overload and short circuit by 63 A BS 88 'gG' fuses. Installation Method 19 is to be used and $t_a = 40$°C.

Determine the minimum conductor cross-sectional area that can be used.

Answer

For cables installed in enclosed trenches the following equation must be satisfied:

$$I_{ta} \times C_a \times C_e \geqslant I_n \text{ A}$$

The relevant table of current-carrying capacity is Table 4D4A. The relevant column in that table is Column 5 because the cables concerned are multicore and Table 4B3 indicates that the current-carrying capacities relating to Installation Method 13 are to be used, that method being limited to multicore cables.

From Table 4C1, $C_a = 0.87$.

Now take a trial value of cross-sectional area – say 16 mm² having $I_{ta} = 83$ A.

From Table 4B3 Column 8, C_e for 16 mm² is 0.71.

Check if the above equation is satisfied, remembering that $I_n = 63$ A.

$$83 \times 0.87 \times 0.71 \text{ A} = 51.3 \text{ A}$$

The equation is *not* satisfied so try 25 mm² having $I_{ta} = 110$ A.

C_e becomes 0.69.

Check:

$$110 \times 0.87 \times 0.69 \text{ A} = 66 \text{ A}$$

As this result *is* greater than I_n the minimum conductor cross-sectional area that can be used is 25 mm².

Note that had the circuits concerned been protected against short circuit only, the equation to be satisfied would then have been:

$$I_{ta} \times C_a \times C_e \geqslant I_b \text{ A}$$

GROUPED CIRCUITS NOT LIABLE TO SIMULTANEOUS OVERLOAD

Inspection of Table 4B1 of Appendix 4 of BS 7671 immediately shows that, particularly for circuits grouped in enclosures or bunched and clipped direct to a non-metallic surface, the grouping factor C_g can lead to a significant increase in the conductor cross-sectional area to be used.

However, for such grouped circuits, BS 7671 in items 6.1.2 and 6.2.2 of the Preface to the tables of conductor current-carrying capacities offers a method of limiting increases in conductor cross-sectional areas – *provided that the grouped circuits are not liable to simultaneous overload.*

This method, when the overcurrent protective devices of the grouped circuits are other than semi-enclosed fuses to BS 3036, consists of calculating I_t from:

$$I_t = \frac{I_b}{C_a C_i C_g} \text{ A}$$

and then calculating I_t from:

$$I_t = \frac{1}{C_a C_i}\sqrt{I_n^2 + 0.48 I_b^2\left(\frac{1 - C_g^2}{C_g^2}\right)} \text{ A}$$

Whichever is the greater of the two values of I_t so obtained is then used when inspecting the appropriate table of conductor current-carrying capacities in order to establish the conductor cross-sectional area having an actual tabulated current-carrying capacity (I_{ta}) equal to or greater than that value of I_t.

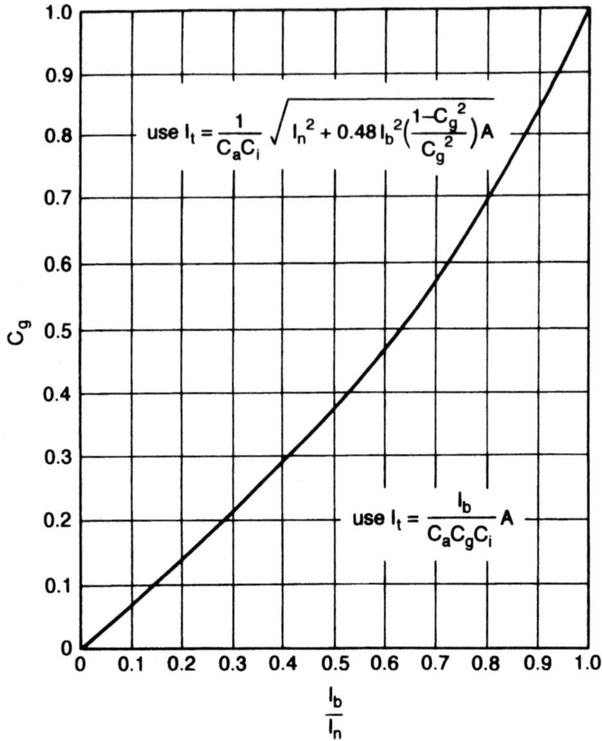

Figure 1.1 Grouped circuits protected by HBC fuses or mcbs and not subject to simultaneous overloads.

Figures 1.1 and 1.2 have been developed as a convenient design aid and are used in the following manner.

For a particular case the values of C_g and I_b/I_n are known and Figure 1.1 immediately shows which of the two expressions gives the higher value of I_t.

Figure 1.2, by using either the broken line (but see comment in Example 1.14) or the appropriate curve from the family of curves, gives the value of the reduction factor F_1. As indicated, this factor F_1 is applied as a multiplier to $I_n/C_aC_gC_i$ to obtain I_t which, as before, is used to determine the minimum conductor cross-sectional area that can be tolerated. Thus Figure 1.2 also gives a very rapid indication as to whether or not it is worthwhile taking advantage of this particular option.

Example 1.14

A single-phase circuit having $I_b = 26$ A is protected by a 32 A BS 88 'gG' fuse and is wired in single-core 70°C pvc-insulated non-sheathed cables to BS 6004 having copper conductors. It is installed with six other similar circuits in conduit on a wall in a location where the expected ambient temperature is 45°C.

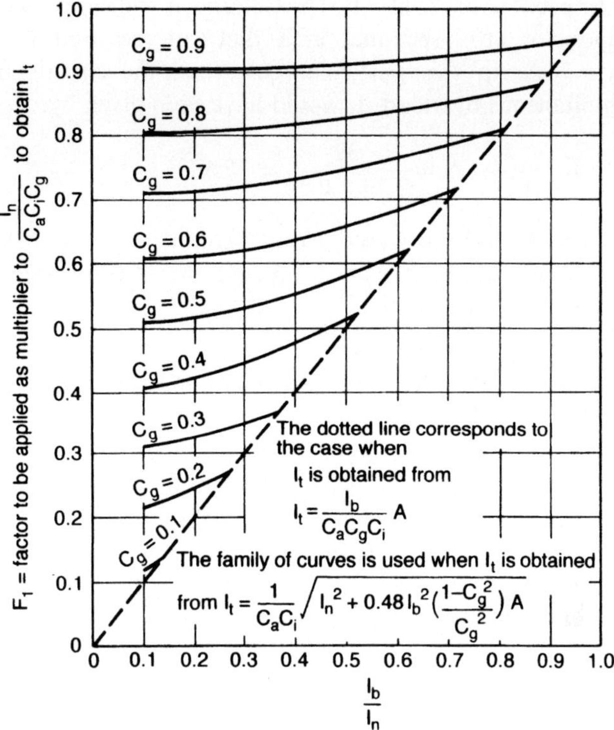

Figure 1.2 Reduction factor to be applied for grouped circuits protected by HBC fuses or mcbs and not subject to simultaneous overloads.

What is the minimum conductor cross-sectional area that can be used if the circuits are not subject to simultaneous overload?

Answer

From Table 4B1, $C_g = 0.54$.

From Table 4C1, $C_a = 0.79$, $C_i = 1$.

$$\frac{I_b}{I_n} = \frac{26}{32} = 0.813$$

From Figure 1.1 for $I_b/I_n = 0.813$ and $C_g = 0.54$ it is found that the expression to use to determine I_t is:

$$I_t = \frac{I_b}{C_aC_g}\,\text{A} = \frac{26}{0.79 \times 0.54}\,\text{A} = 60.9\,\text{A}$$

Note that in such cases there is no point in using the dotted line in Figure 1.2 in order to determine the 'reduction factor' because this is simply the ratio I_b/I_n.

Inspection of Table 4D1A Column 4 indicates that the minimum conductor cross-sectional area that can be used is 16 mm^2 having $I_{ta} = 76$ A. In this particular case, had the circuits been subject to simultaneous overload, I_t would have been given by:

$$I_t = \frac{I_n}{C_a C_g}\,A = \frac{32}{0.79 \times 0.54} = 75\,A$$

One could still have used 16 mm^2 conductor area so that, in fact, no advantage has been gained because the circuits were not subject to simultaneous overload.

Example 1.15

Eight similar three-phase circuits are wired in 85°C rubber-insulated multicore cables to BS 6883 having copper conductors, these being bunched and clipped direct, $t_a = 50°C$, $I_b = 34$ A and $I_n = 40$ A, the protection against overcurrent being provided by miniature circuit breakers.

If it can be assumed that the circuits are not subject to simultaneous overload what is the minimum conductor cross-sectional area that can be used?

Answer

From Table 4B1, $C_g = 0.52$.

From Table 4C1, $C_a = 0.80$, $C_i = 1$.

$$\frac{I_b}{I_n} = \frac{34}{40} = 0.85$$

For that value of I_b/I_n and $C_g = 0.52$, from Figure 1.1 it is found that the expression to determine I_t is:

$$I_t = \frac{I_b}{C_a C_g}\,A = \frac{34}{0.80 \times 0.52}\,A = 81.7\,A$$

From Table 4F2A Column 3 it is then found that the minimum conductor cross-sectional area to use is 16 mm^2 having $I_{ta} = 88$ A.

In this case some advantage has been gained because if one assumed that the circuits were subject to simultaneous overload I_t would have been given by:

$$I_t = \frac{I_n}{C_a C_g}\,A = \frac{40}{0.80 \times 0.52}\,A = 96.2\,A$$

and it would have been found that the minimum conductor cross-sectional area would be 25 mm^2.

For grouped circuits protected by semi-enclosed fuses to BS 3036 which are not liable to simultaneous overload, a similar method to that just described can be used, except that one of the equations is different.

One first has to calculate. as before, I_t from:

$$I_t = \frac{I_b}{C_a C_i C_g} \text{ A}$$

and then from:

$$I_t = \frac{1}{C_a C_i}\sqrt{1.9 I_n^2 + 0.48 I_b^2\left(\frac{1 - C_g^2}{C_g^2}\right)} \text{ A}$$

Whichever is the greater of the two values so obtained is then used when inspecting the appropriate table of conductor current-carrying capacities in order to establish the conductor cross-sectional area having an actual tabulated current-carrying capacity (I_{ta}) equal to or greater than that value of I_t.

Figures 1.3 and 1.4 have been developed as a convenient design aid, similar to Figures 1.1 and 1.2 but for circuits protected by semi-enclosed fuses.

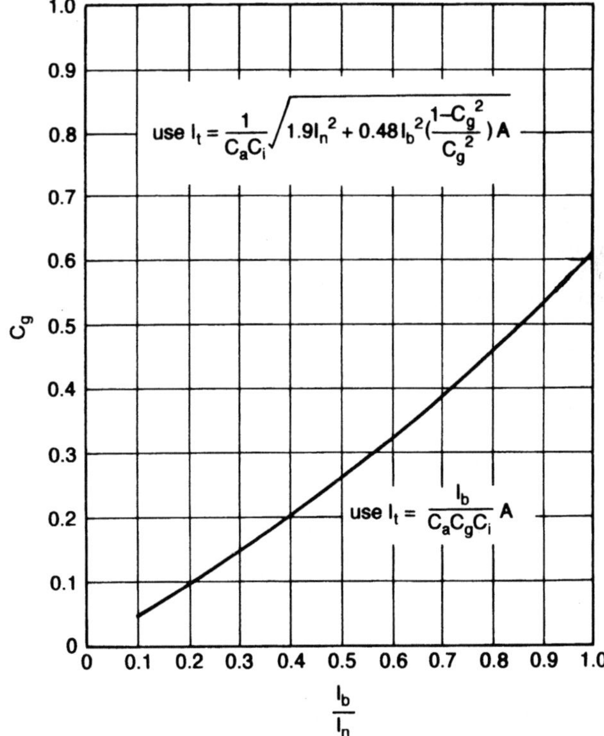

Figure 1.3 Grouped circuits protected by semi-enclosed fuses to BS 3036 and not subject to simultaneous overloads.

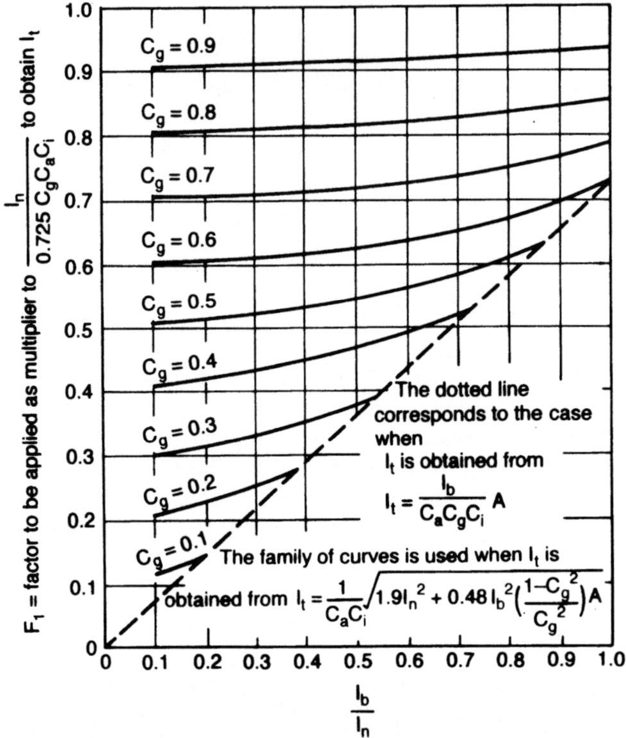

Figure 1.4 Reduction factor to be applied for grouped circuits protected by semi-enclosed fuses to BS 3036 and not subject to simultaneous overloads.

Example 1.16

Seven similar single-phase circuits having $I_b = 25$ A are each protected against both overload and short circuit by a 30 A BS 3036 semi-enclosed fuse. Two-core 70°C pvc-insulated and sheathed non-armoured cables to BS 6004 are used, the circuits being bunched and clipped direct to a wall and $t_a = 35$°C.

What is the minimum conductor cross-sectional area that can be used if it can be assumed that the circuits are not liable to simultaneous overload?

Answer

From Table 4B1, $C_g = 0.54$.

From Table 4C2, $C_a = 0.97$, $C_i = 1$.

$$\frac{I_b}{I_n} = \frac{25}{30} = 0.83$$

From Figure 1.3 for $I_b/I_n = 0.83$ and $C_g = 0.54$ it is seen that I_t has to be found from:

$$I_t = \frac{1}{C_a C_i}\sqrt{1.9I_n^2 + 0.48I_b^2\left(\frac{1 - C_g^2}{C_g^2}\right)}\ \text{A}$$

From Figure 1.4 it is then found that the 'reduction factor' F_1 is approximately 0.65, so that:

$$I_t = \frac{F_1 \times I_n}{0.725 \times C_g \times C_a}\ \text{A} = \frac{0.65 \times 30}{0.725 \times 0.54 \times 0.97}\ \text{A} = 51.3\,\text{A}$$

From Table 4D2A Column 6 the minimum conductor cross-sectional area that can be used is found to be $10\,\text{mm}^2$ having $I_{ta} = 63\,\text{A}$.

In those cases where there is very little difference between I_t and I_{ta} it would be advisable to use the actual formula as a check. In the present case there is no need to do this but I_t so obtained would have been found to be 50.9 A.

BS 7671 offers another way of effecting some reduction in the cross-sectional areas of grouped cables as indicated in Note 2 to Tables 4B1 and 4B2.

Normally a group of N loaded cables require a grouping correction factor of C_g applied to the actual tabulated current-carrying capacity (I_{ta}). However, if M cables of that group are known to be carrying currents not greater than $0.3C_gI_{ta}$ A the other cables can be sized by using the grouping correction factor corresponding to (N – M) cables.

CIRCUITS IN LOW AMBIENT TEMPERATURES

All the examples so far, except one, have concerned an ambient temperature of 30°C or greater. Consider now the case where the expected ambient temperature is less than 30°C (i.e. is less than the reference ambient temperature).

Table 4C1 of BS 7671 gives ambient temperature correction factors for overcurrent protective devices other than semi-enclosed fuses to BS 3036 and principally for ambient temperatures exceeding 30°C. These factors are based on the fact that with such ambient temperatures the limiting parameter is the maximum normal operating temperature for the type of cable insulation concerned and are given by:

$$C_a = \sqrt{\left(\frac{t_p - t_a}{t_p - 30}\right)}$$

These factors therefore apply if the protective device is providing overload protection with or without short circuit protection or is providing short circuit protection only.

However, for ambient temperatures below 30°C it will be found that the limiting parameter is now the maximum tolerable temperature under overload conditions (t_o) where:

$$t_o = t_r + 1.45^2(t_p - t_r)°\text{C}$$

and C_a is then given by:

$$C_a = \frac{1}{1.45}\sqrt{\left(\frac{t_o - t_a}{t_p - 30}\right)}$$

It follows from the foregoing that at these low ambient temperatures the ambient temperature correction factors to be used when the overcurrent protective device is providing overload protection will be different to those if it is providing only short circuit protection.

Table 1.2 gives the ambient temperature correction factors for protective devices other than semi-enclosed fuses.

Table 1.2 Correction factors for low ambient temperatures.

Type of cable insulation	Function of overcurrent protective device	Ambient temperature (°C)					
		0	5	10	15	20	25
70°C pvc	Overload, with or without short circuit protection	1.17	1.14	1.12	1.09	1.06	1.03
	Short circuit protection only	1.32	1.27	1.22	1.17	1.12	1.06
85°C rubber	Overload, with or without short circuit protection	1.12	1.10	1.08	1.06	1.04	1.02
	Short circuit protection only	1.24	1.21	1.17	1.13	1.08	1.04
XLPE (i.e. 90°C thermosetting)	Overload, with or without short circuit protection	1.11	1.09	1.07	1.05	1.04	1.02
	Short circuit protection only	1.22	1.19	1.15	1.12	1.08	1.04

The ambient temperature correction factors given in Table 1.2 are used in exactly the same way as those given in Table 4C1 of BS 7671.

Throughout this book the term 'XLPE insulation' is used, XLPE standing for cross-linked polyethylene. Currently this is the most commonly used type of 90°C thermosetting insulation. Other compounds, when developed, will not necessarily have the same characteristics (such as the k value) as XLPE.

Where XLPE insulated cables are connected to equipment or accessories designed for a maximum temperature of 70°C the current-current carrying capacity of an equivalent pvc cable should be used for the XLPE cable. Where this leads to a larger conductor size being selected the actual operating temperature at the terminals will be lower. Advantage can still be taken of the higher overload and short circuit capacity of the XLPE insulated cable.

Example 1.17

Four similar three-phase circuits are run in multicore 70°C pvc-insulated non-armoured cables having aluminium conductors laid in flush floor trunking (Installation Method 9).

If each circuit is protected against both overload and short circuit by 63 A BS 88 'gG' fuses and $t_a = 10°C$, what is the minimum conductor cross-sectional area that can be used?

Answer

From Table 4A the *Reference* Method is Method 3.

From Table 4B1, $C_g = 0.65$.

From Table 1.2 above, $C_a = 1.12$.

Thus:

$$I_t = 63 \times \frac{1}{0.65} \times \frac{1}{1.12} \text{ A} = 86.5 \text{ A}$$

From Table 4K2A Column 5 the minimum conductor cross-sectional area that can be used is found to be 50 mm^2, having $I_{ta} = 92$ A.

GROUPED RING CIRCUITS

A problem frequently encountered in practice is that of grouped ring circuits.

For such a case:

$$I_{ta} \geqslant I_n \times \frac{1}{C_r} \times \frac{1}{C_a} \times \text{A}$$

where C_a is the ambient temperature correction factor and
C_r is the ring circuit grouping factor.

If, as may happen, the grouped circuits are embedded in thermal insulation the factor C_i also applies and then:

$$I_{ta} \geqslant I_n \times \frac{1}{C_r} \times \frac{1}{C_a} \times \frac{1}{C_i} \text{ A}$$

It should be noted that the equation applies equally to circuits protected by BS 3036 semi-enclosed fuses and to those protected by HBC fuses or mcbs.

On the assumption that in each of the grouped circuits there is a 67 : 33 split in the total load current, the values of C_r given in Table 1.3 apply.

Table 1.3 Correction factors for grouped ring circuits.

Number of ring circuits	1	2	3	4	5	6	7	8	9	10
C_r	1.5	1.19	1.03	0.94	0.87	0.82	0.77	0.74	0.71	0.69

Example 1.18

Eight single-phase ring circuits are grouped in a common trunking. It is intended to use two-core (with cpc) 70°C pvc-insulated and sheathed flat cables to BS 6004 having copper conductors.

If $t_a = 40°C$ and each circuit is to be protected by a 30 A mcb, what is the minimum conductor cross-sectional area that can be used?

Answer

From Table 1.3 above, $C_r = 0.74$.

From Table 4C1, $C_a = 0.87$.

Thus:

$$I_t = 30 \times \frac{1}{0.74} \times \frac{1}{0.87}\ \text{A} = 46.6\ \text{A}$$

Because Table 4D5A does not include current-carrying capacities for flat twin and earth cables in trunking Table 4D2A has to be consulted. From Column 4 of this table it is found that the minimum conductor cross-sectional area that can be used is $10\ \text{mm}^2$ having $I_{ta} = 52\ \text{A}$.

MOTOR CIRCUITS SUBJECT TO FREQUENT STOPPING AND STARTING

Regulation 552–01–01 requires that where a motor is intended for intermittent duty and for frequent stopping and starting, account shall be taken of any cumulative effects of the starting periods upon the temperature rise of the equipment of the circuit. That equipment, of course, includes the cables feeding the motor.

When motors are regularly restarted soon after they have been stopped the cables supplying the motor may not have time to cool to close to the ambient temperature, this time being anything from 20 minutes or so to several hours depending on their size and method of installation.

Furthermore starting currents (for direct-on-line starting) can be between 5 and 8 times the full load current.

The calculation of any accurate factor to take account of frequent stopping and starting is complex and requires that information is provided on:

(a) the actual magnitude of the starting current
(b) the duration of, and rate of decay, of the starting current
(c) the full load current
(d) the first choice cable size to calculate its time constant
(e) the time intervals between stopping and starting.

It is *suggested* that the cable cross-sectional area is chosen on the basis of its current-carrying capacity being 1.4 times the full load current.

Example 1.19

A 10 hp 240 V single-phase motor is subject to frequent stopping and starting. The cable supplying the motor via a direct-on-line starter is two-core, non-armoured cable having 85°C rubber insulation and copper conductors, clipped direct.

If the ambient temperature is 50°C what should be the conductor cross-sectional area?

Answer

As 1 hp = 746 watts the full load current is:

$$\frac{10 \times 746}{240} \text{ A} = 31.1 \text{ A}$$

The conductor current-carrying capacity should therefore be at least 1.4 × 31.1 A = 43.5 A.

As the ambient temperature is 50°C, from Table 4C1 (assuming the circuit overcurrent protective device is other than a BS 3036 semi-enclosed fuse) C_a is found to be 0.80.

The cross-sectional area of the conductor should therefore have I_{ta} of at least 43.5/0.8 A = 54.4 A.

From Table 4F2A Column 4 it is found that the minimum conductor cross-sectional area to use is 6 mm^2 having I_{ta} = 55 A.

Due account has to be taken of the motor efficiency and motor power factor where these are known.

Example 1.20

A 415 V three-phase motor has a rating of 15 kW at a power factor of 0.8 lagging, the efficiency being 90%.

If this motor is subject to frequent stopping and starting and it is intended to use a multicore armoured 70°C pvc-insulated cable having copper conductors, installed in free air, what should be their cross-sectional area when t_a = 35°C?

Answer

The full load current of the motor is:

$$\frac{15 \times 1000}{0.8 \times 0.9 \times \sqrt{3} \times 415} \text{ A} = 29 \text{ A (per phase)}$$

The conductor current-carrying capacity should therefore be at least 1.4 × 29 A = 40.6 A.

As the ambient temperature is 35°C, from Table 4C1 (assuming the circuit overcurrent protection device is other than a semi-enclosed fuse to BS 3036) C_a is found to be 0.94.

The cross-sectional area of the conductors should therefore have I_{ta} of at least 40.6/0.94 A = 43.2 A.

From Table 4D4A Column 5 it is found that the minimum conductor cross-sectional area is 6 mm^2 having $I_{ta} = 45$ A.

It will be noted that in this example no mention is made of the nominal current rating of the overcurrent protective device for the motor circuit. The omission is deliberate because that device, in very many motor circuits, is providing short circuit protection only. Overload protection is provided by, for example, thermal overload relays in the motor starter. The nominal current rating of the overcurrent protective device in the motor circuit and its operating characteristics must be such as to provide adequate short circuit protection to the circuit conductors (i.e. give compliance with the adiabatic equation in Regulation 434–03–03) but not cause disconnection of the motor circuit because of high starting currents.

CIRCUITS FOR STAR-DELTA STARTING OF MOTORS

Another problem associated with motors is the current demand when star-delta starting is used. With star-delta starters six conductors (two per phase) are required between the starter and the motor, the star or delta connections being made in the starter.

During the running condition the current carried by each of the six cables is $1/\sqrt{3}$ times that of the supply cables to the starter (the connection being delta). If the motor cables are not grouped they require a current-carrying capacity 58% that of the supply cables. In practice it is far more likely that the motor cables are grouped together so that the grouping correction factor (C_g) for two circuits applies and the required current-carrying capacity for the motor cables becomes 72% that for the supply cables assuming the cables are either enclosed or bunched and clipped to a non-metallic surface.

If a delta-connected motor is started direct-on-line the starting current may be between 5 and 8 times the full load current depending on the size and design of that motor. When star-delta starting is used the current taken from the supply is one third of that taken when starting direct on line. Hence the starting current in star may be between 1.67 and 2.67 times the full load current. During the starting period, because the star connection is used, the current in each of the motor cables equals that in the supply cables.

Because the starting current is only carried for a short period advantage can be taken of the thermal inertia of the cables and the cables chosen on the basis of their running condition will be satisfactory for the starting current.

Example 1.21

A 415 V three-phase 20 kW motor has a power factor of 0.86 lagging and an efficiency of 90%. Star-delta starting is to be used. The supply cables to the starter are single-core 70°C pvc-insulated (copper conductors) in conduit. The motor cables from the starter are also single-core pvc-insulated (copper conductors) and in another length of conduit.

What is the minimum conductor cross-sectional area for both circuits if $t_a = 30°C$?

Answer

$$\text{The line current} = \frac{20 \times 1000}{\sqrt{3} \times 415 \times 0.86 \times 0.90} \text{ A} = 35.9 \text{ A}$$

There are no correction factors to be applied.

From Table 4D1A Column 5, it is found that the minimum conductor cross-sectional area for the supply cables is 6 mm² having $I_{ta} = 36$ A.

For the cables from the starter to the motor, each cable carries

$$\frac{35.9}{\sqrt{3}} \text{ A} = 20.7 \text{ A}$$

But now it is necessary to apply the grouping factor for two circuits which from Table 4B1 is 0.8.

Thus:

$$I_t = \frac{20.7}{0.8} \text{ A} = 25.9 \text{ A}$$

Again from Table 4D1A Column 5 it is found that these cables can be of 4 mm² cross-sectional area, having $I_{ta} = 28$ A.

As already stated, whilst during starting the cables between the motor and starter may be carrying a current of 35.9 × 2.67 A (=96 A), but the time the motor is held in star will generally be less than the time taken for these cables to reach the maximum permitted normal operating temperature.

CHANGE OF PARAMETERS OF ALREADY INSTALLED CIRCUITS

The occasion may arise, for example when there is a change of tenancy of the premises concerned or it is intended to change the load or other circuit details, that the user wishes to determine whether an already installed cable will be adequate. In such cases the following method should be used.

The type of cable and its cross-sectional area are established so that the relevant table in Appendix 4 of BS 7671 and the appropriate column in that table are identified. From that table and column, I_{ta} is obtained.

Then determine the product:

$$I_{ta} \times C_a \times C_g \times C_i \text{ A}$$

if the protective device it is intended to use is other than a semi-enclosed fuse to BS 3036. Denote this product by I_{tc} A.

The cable concerned could therefore be used:

(a) for a circuit protected against both overload and short circuit by an HBC fuse or an mcb having a nominal current rating not greater than I_{tc} A, and when the design current (I_b) of the circuit is less than that nominal current rating, *or*
(b) for a circuit only protected against short circuit by an HBC fuse or mcb where the design current of the circuit is not greater than I_{tc} A. The nominal current rating of the device can be greater than I_{tc} (see Regulation 434–03–01) and is determined by the load characteristics together with the need to comply with Regulation 434–03–03.

If it is intended to use semi-enclosed fuses to BS 3036 then:

$$I_{tc} = I_{ta} \times C_a \times C_g \times C_i \times 0.725\,A$$

but *only* if it is intended that the fuses are to give overload protection, with or without short circuit protection. In the unlikely event that such a fuse is intended to give only short circuit protection then the 0.725 factor is *not* used to determine I_{tc}.

The symbol I_{tc} has been used here in an attempt to clarify a point which sometimes causes confusion. When the overcurrent protective device is other than a semi-enclosed fuse to BS 3036, I_{tc} has exactly the same identity as what is termed in BS 7671 the current-carrying capacity for continuous service under the particular installation conditions concerned and denoted by I_z.

When the overcurrent protection device is a semi-enclosed fuse to BS 3036, I_z is given by:

$$I_z = I_{ta} \times C_a \times C_g \times C_i\,A$$

i.e. exactly the same expression as that for HBC fuses and mcbs. In other words, the effective current-carrying capacity of a conductor is totally independent of the type of overcurrent protective device associated with that conductor.

The factor 0.725 is only applied when determining the nominal current of a semi-enclosed fuse that can be used for a given conductor in order to take account of the higher fusing factor of such a fuse.

Example 1.22

An existing cable for a three-phase circuit is multicore non-armoured 70°C pvc-insulated having copper conductors of 25 mm^2 cross-sectional area and is installed with five other similar cables in trunking.

What is the maximum nominal current of 'gG' fuses to BS 88 Pt 2 that can be used to protect this cable against both overload and short circuit if $t_a = 50°C$?

Answer

From Table 4D2A Column 5, $I_{ta} = 80\,A$.

From Table 4B1, $C_g = 0.57$ (there being a total of six circuits).

From Table 4C1, $C_a = 0.71$.

As $C_i = 1$

$$I_{tc} = 80 \times 0.57 \times 0.71\,A = 32.4\,A$$

From the standard nominal current ratings for BS 88 Pt 2 fuses the maximum that can be used is 32 A.

The design current I_b also must not exceed 32 A.

Example 1.23

An existing cable for a three-phase circuit is multicore armoured having XLPE insulation and copper conductors of 35 mm^2 cross-sectional area. It is clipped direct and is not bunched with any other cable.

If $t_a = 10°C$ and it is intended to protect the cable against short circuit current only, using HBC fuses, what is the maximum design current that can be tolerated?

Answer

From Table 1.2 of this Chapter $C_a = 1.15$.

Also $C_g = 1$ and $C_i = 1$.

From Table 4E4A, Column 3, $I_{ta} = 154\,A$.

Then:

$$I_{tc} = 154 \times 1.15\,A = 177\,A$$

The maximum value of I_b therefore is 177 A.

The nominal current rating of the HBC fuse may be greater than 177 A. It will be determined by the load characteristics and will possibly be limited by the need to comply with Regulation 434–03–03.

Example 1.24

An existing cable is a two-core flat 70°C pvc-insulated cable having copper conductors of 16 mm^2 which is clipped direct, bunched with four other similar cables.

If $t_a = 40°C$ and it is intended to protect the cable using a BS 3036 semi-enclosed fuse against overload, what is the maximum nominal rating that fuse can have?

Answer

From Table 4D5A, Column 4, $T_{ta} = 85\,A$.

From Table 4B1, $C_g = 0.60$.

From Table 4C2, $C_a = 0.94$.

Thus:

$$I_{tc} = 85 \times 0.60 \times 0.94 \times 0.725\,A = 34.8\,A$$

The maximum nominal current rating of the semi-enclosed fuse is therefore 30 A.

The design current of the circuit also must not exceed 30 A.

Example 1.25

An existing cable is an armoured multicore cable having aluminium conductors of 70 mm^2 cross-sectional area and XLPE insulation. It is clipped direct to a non-metallic surface, not grouped with the cables of other circuits and $t_a = 50°C$. The three-phase equipment fed by this cable is to be changed for equipment having a much higher kW rating and it is proposed to run a similar cable in parallel with that existing.

If overcurrent protection (both overload and short circuit) is to be provided by BS 88 'gG' fuses, what is the maximum nominal rating of those fuses and the maximum load current for the circuit?

Answer

From Table 4L4A, Column 3, $I_{ta} = 174\,A$.

From Table 4C1, $C_a = 0.82$.

Assuming that the two cables will be touching or very close together it is also necessary to apply the appropriate grouping factor, C_g.

From Table 4B1, $C_g = 0.8$.

Thus:

$$I_{tc} = 174 \times 0.82 \times 0.8\,A = 114\,A$$

As there are two cables in parallel this value has to be doubled i.e. $I_{tc} = 228\,A$.

From the standard nominal ratings for BS 88 'gG' fuses it is found that the maximum that can be used is 200 A and this is also the maximum value of I_b that can be tolerated for the circuit.

ADMIXTURES OF CABLE SIZES IN ENCLOSURES

Last but far from least in this chapter the problem of mixed cable sizes and loads in trunking or other cable enclosures is considered.

The table of grouping correction factors (Table 4B1) in BS 7671, as indicated in Note 1 to that table, is applicable only to groups of cables where the cables are all of one size and are equally loaded. For this reason all the examples so far given in this chapter have concerned such groups but in practice it is infinitely more likely that trunking or conduit will contain cables of various sizes and with different loadings.

Where the range of sizes in a particular enclosure is not great, using the appropriate grouping correction factor from Table 4B1 can be justified. However, where a wide range of sizes is accommodated in an enclosure, using the Table 4B1 correction factor can lead to oversizing the large cables but undersizing of the small cables.

The following method is based on one that was developed by Mr N.S. Bryant when he was with Pirelli General plc and put forward by him at an IEE discussion meeting in March 1984. There are six steps in the method and, for the purpose of illustration, after the description of each step the values for the specific example chosen are given.

Example 1.26

It is intended to run in 75 mm × 50 mm trunking the following circuits:

Circuit 1	Three-phase circuit $I_b = 60\,A$
Circuit 2	Three-phase circuit $I_b = 50\,A$
Circuit 3	Single-phase circuit $I_b = 40\,A$
Circuit 4	Single-phase circuit $I_b = 30\,A$
Circuit 5	Single-phase circuit $I_b = 5\,A$
Circuit 6	Single-phase circuit $I_b = 5\,A$
Circuit 7	Single-phase circuit $I_b = 5\,A$

The expected ambient temperature is 30°C. Determine the conductor cross-sectional areas, using for all the circuits single-core 70°C pvc-insulated cables having copper conductors and check they are thermally adequate.

Answer

Step 1

Establish the conductor cross-sectional areas based on voltage drop requirements or by experience or by applying a single grouping correction factor.

Assume it is decided to use the following cables:

Circuit 1	3 cables	$35\,mm^2$
Circuit 2	3 cables	$35\,mm^2$
Circuit 3	2 cables	$25\,mm^2$
Circuit 4	2 cables	$25\,mm^2$
Circuit 5	2 cables	$1.5\,mm^2$
Circuit 6	2 cables	$1.5\,mm^2$
Circuit 7	2 cables	$1.5\,mm^2$

Step 2

Calculate the 'fill factor' for the chosen sizes of cable and trunking where:

$$\text{fill factor} = \frac{\text{sum of the overall cable cross-sectional areas}}{\text{internal area of the trunking}}$$

The cable cross-sectional areas are found from the cable manufacturer's data either as such or, if a cable overall diameter (D_e) is given, by:

$$\frac{\pi}{4} D_e^2 \text{ mm}^2$$

For the example the cable cross-sectional areas are:

for $35\,\text{mm}^2$–$95\,\text{mm}^2$
for $25\,\text{mm}^2$–$75.4\,\text{mm}^2$
for $1.5\,\text{mm}^2$–$9.1\,\text{mm}^2$

The sum of all the cable cross-sectional areas is:

$$(6 \times 95) + (4 \times 75.4) + (6 \times 9.1)\,\text{mm}^2 = 926\,\text{mm}^2$$

The internal area of the trunking is $(75 \times 50)\,\text{mm}^2 = 3750\,\text{mm}^2$

$$\text{The fill factor} = \frac{926}{3750} = 0.247$$

Note that the fill factor is *not* the same as the space factor which is used to determine whether a cable enclosure can accommodate the desired number of cables. When determining the fill factor the existence of circuit protective conductors and, in the case of balanced three-phase circuits, the neutral conductor, is ignored because they do not normally carry current.

It should be noted that examination of steps 2, 3 and 5 reveals that it is not necessary to calculate the trunking fill factor but only to determine the area occupied by the cables. However, the advantage in calculating the fill factor is that it is a convenient value to use when tabulating values of thermal capability. If this method is to be used frequently values of thermal capability can be calculated and tabulated for a number of different fill factors and retained for future use.

Step 3

Find the number of cables (N) of any particular size ($S\,\text{mm}^2$) which will give the same fill factor as that determined in Step 2.

It will be found that 26 cables of $10\,\text{mm}^2$ conductor cross-sectional area (overall area being $36.3\,\text{mm}^2$) will give a fill factor of 0.252 which is sufficiently close to 0.247.

Step 4

Calculate the thermal capability of the trunking for the fill factor, the thermal capability being given by:

$$(I_{ta} \times C_g)^2 \times N \times R \text{ watts/metre}$$

where: I_{ta} = current-carrying capacity for the conductor cross-sectional area chosen in Step 3
C_g = grouping correction factor for the number of cables chosen in Step 3, i.e. for N cables
R = effective a.c. resistance per cable per metre in ohms/m.

Although some circuits are three-phase and others are single-phase, to find C_g for the 26 cables, take from Table 4B1 the value for 26/2, i.e. 13 circuits (single-phase). This value is namely 0.44.

Similarly to find I_{ta} use Column 4 of Table 4D1A (i.e. the column for single-phase circuits). It is found to be 57 A for 10 mm^2 cross-sectional area.

A very convenient way of finding R is to use:

$$\frac{\text{tabulated mV/A/m}}{1000 \times 2} \text{ ohms/m}$$

where the tabulated mV/A/m relates to single-phase circuits. From Table 4D1B it is found to be 4.4 milliohms/m for 10 mm^2 cross-sectional area.

Thus the thermal capability of the trunking is:

$$\frac{(57 \times 0.44)^2 \times 26 \times 4.4}{1000 \times 2} \text{ W/m} = 36 \text{ W/m}$$

Step 5

Calculate for each actual cable size in the trunking its permitted power dissipation (P) which is given by:

$$\frac{\text{its own overall c.s.a.} \times \text{thermal capacity of trunking}}{\text{fill factor} \times \text{internal area of the trunking}} \text{ W/m}$$

So that:

for 35 mm^2 $\quad P = \dfrac{95 \times 36}{0.252 \times 3750} \text{ W/m} = 3.62 \text{ W/m}$

for 25 mm^2 $\quad P = \dfrac{75.4 \times 36}{0.252 \times 3750} \text{ W/m} = 2.87 \text{ W/m}$

for 1.5 mm^2 $\quad P = \dfrac{9.1 \times 36}{0.252 \times 3750} \text{ W/m} = 0.347 \text{ W/m}$

Step 6

Calculate for each actual cable size in the trunking its grouped current-carrying capacity (I_g) which is given by:

$$I_g \sqrt{\frac{P}{R}} A$$

where: P is its power dissipation as calculated in Step 5
R is its resistance in ohms/m.

R is determined as before:

$$R = \frac{\text{tabulated mV/A/m}}{2 \times 1000} \text{ ohms/m}$$

using the tabulated mV/A/m for single-phase circuits, or:

$$R = \frac{\text{tabulated mV/A/m}}{\sqrt{3} \times 1000} \text{ ohms/m}$$

using the tabulated mV/A/m for three-phase circuits. (For a given conductor size the R values so obtained are the same.) So that using mV/A/m values from Table 4DIB:

for $35\,mm^2$ $R = \frac{1.3}{2000}$ ohms/m = 0.00065 ohms/m

for $25\,mm^2$ $R = \frac{1.8}{2000}$ ohms/m = 0.00090 ohms/m

for $1.5\,mm^2$ $R = \frac{29}{2000}$ ohms/m = 0.0145 ohm/m

and, therefore the values of I_g are:

for circuit 1	74.6 A
for circuit 2	74.6 A
for circuit 3	56.5 A
for circuit 4	56.5 A
for circuit 5	4.9 A
for circuit 6	4.9 A
for circuit 7	4.9 A

If these values of I_g are compared with the design currents, I_b, for the circuits it will be seen that for circuits 1, 2, 3 and 4 they are greater whereas for circuits 5, 6 and 7 they are marginally lower.

It is a matter of personal judgement of the designer to attempt to change the proposed conductor cross-sectional area for one or more of the circuits. If it is decided to follow this course it is necessary to repeat the above procedure.

On the other hand because there is some 'spare' current-carrying capacity in some of the circuits it is possible to trade off this against the undersizing of the three 5 A circuits.

When the total power dissipation per metre from all the circuits is less than the thermal capacity of the trunking per metre a cable may exceed its grouping current carrying capacity by a factor F_g which is given by

$$F_g = \sqrt{3.95 - (0.0295 \times \mu)}$$

$$\text{where: } \quad \mu = \frac{\text{total power dissipation of the cables per metre} \times 100}{\text{thermal capacity of the trunking per metre}}$$

In the present case the various actual power dissipations per cable (I_b^2R), calculated using the I_b values given earlier and R values from Step 6, are given below. The total (W/m) values relate to the circuits and are the per cable values multiplied by 3 for three-phase circuits and by 2 for single-phase circuits.

	per cable (W/m)	total (W/m)
Circuit 1	2.34	7.02
Circuit 2	1.63	4.88
Circuit 3	1.44	2.88
Circuit 4	0.81	1.62
Circuit 5	0.363	0.726
Circuit 6	0.363	0.726
Circuit 7	0.363	0.726

The total cable power dissipation per metre is therefore 18.6 W/m, this is less than the thermal capacity of the trunking.

$$\mu = \frac{18.6}{36} \times 100\% = 51.7\%$$

$$F_g = \sqrt{3.95 - (0.0295 \times 51.7)} = 1.56$$

Hence the maximum current which could be carried by the light current circuits is $1.56 \times 4.9\,\text{A} = 7.6\,\text{A}$.

The above example is a relatively simple one and readers wishing to have more detailed information should study ERA Reports 69–30 Parts VIII and IX in which much of the required data has been tabulated.

In those cases where the circuits concerned are subject to overload, having used the above method to determine the correct conductor cross-sectional areas from consideration of current-carrying capacity it is then necessary to check these areas to ensure that the circuits are adequately protected against overload.

To do this the methods given earlier in this chapter are used and it becomes necessary to determine the grouping factor (C_g) for each circuit.

This is given by:

$$C_g = \frac{I_g \times F_g}{I_{ta}}$$

where, as before: I_g = group current-carrying capacity, A (see Step 6)
F_g = factor (again see Step 6)
I_{ta} = tabulated current-carrying capacity, A for a single circuit (in trunking), for the cable concerned.

GROUPING OF CABLES HAVING DIFFERENT INSULATION

In practice the situation may arise where cables are grouped which are of different types having different maximum permitted operating temperatures. In such cases further correction factors have to be applied to the cables having the higher maximum permitted operating temperatures so that all the cable ratings are based on the lowest such temperature of any cable in the group.

These correction factors are given in Table 1.4. For example, if a group contains 70°C pvc-insulated cables, 85°C rubber insulated cables and cables having XLPE insulation, correction factors of 0.85 and 0.82 should be applied to the ratings of the last two named respectively in addition to the conventional grouping factors.

Alternatively the current ratings for all of the cables in the group can be selected from the table appropriate to the cable having the lowest maximum permitted operating temperature.

Table 1.4 Correction factors for groups of different types of cable.

Conductor maximum permitted operating temperature	Maximum permitted operating temperature in the group			
	60°C	70°C	85°C	90°C
60°C	1	—	—	—
70°C	0.87	1	—	—
85°C	0.74	0.85	1	—
90°C	0.71	0.82	0.96	1

Chapter 2
Calculation of Voltage Drop Under Normal Load Conditions

The designer will frequently find that compliance with the voltage drop requirements is the predominant factor in determining the minimum conductor cross-sectional area that can be used for a particular circuit.

When calculating the voltage drop of a circuit the designer has the option of using the 'simple approach' or, in many cases, a more accurate design approach.

The first series of examples in this chapter illustrates the simple approach which uses the tabulated mV/A/m values given in Appendix 4 of BS 7671 directly, i.e. without taking account of conductor operating temperature or load power factor. This simple approach is a pessimistic one which sometimes leads to larger conductor cross-sectional areas than are necessary.

The second series of examples illustrates the more accurate design approach and some indication is given of where this may be used to advantage.

In general, when determining voltage drop the designer should initially use the simplified approach and only when the voltage drop indicated by this approach exceeds the desired value should the designer consider using the more accurate method described later. However, in most cases where the value obtained from the simplified approach only marginally exceeds the desired value the designer should always recalculate using the more accurate method as it may be found that this may obviate the need to increase the conductor cross-sectional area.

In some of the examples of this chapter a comparatively high ambient temperature is quoted. This has been done simply for the purpose of illustration and it must be emphasised that in the great majority of practical cases the designer is able to assume that the expected ambient temperature will be either the reference value (30°C) or near to it.

THE SIMPLE APPROACH

The only information needed is:

- type of cable
- the conductor cross-sectional area
- the method of installation (for a.c. circuits only)
- the circuit route length (l)
- the type of circuit (d.c., single-phase a.c. or three-phase a.c.)
- the load on the circuit.

The following factors are *not* needed:

(i) the type and nominal current rating of the associated overcurrent protective device
(ii) the ambient temperature
(iii) whether the circuit is run singly or grouped with other circuits
(iv) the power factor of the load.

For d.c. circuits using conductors of any cross-sectional area, and for a.c. circuits using conductors of 16 mm² or less cross-sectional area:

$$\text{Voltage drop} = \frac{\text{tabulated mV/A/m} \times I_b \times 1}{1000} \text{volts}$$

For a.c. circuits using conductors of 25 mm² or greater cross-sectional area,

$$\text{Voltage drop} = \frac{\text{tabulated (mV/A/m)}_z \times I_b \times 1}{1000} \text{volts}$$

Note that the tabulated $(\text{mV/A/m})_z$ value will be found in the column appropriate to the method of installation and type of circuit but in the sub-column headed 'Z'.

It should also be noted that in all cases it is admissible to calculate the voltage drop using I_b and not I_n.

In the voltage drop tables of Appendix 4 of BS 7671 the heading used is 'Voltage drop per ampere per metre' and the tabulated values are given in millivolts. This approach does not lead to any misunderstanding and, as indicated by the two formulae above, one can readily determine the voltage drop of whatever type of circuit, using the appropriate tabulated value.

However, it can be argued that these tabulated mV/A/m values are strictly in milliohms/m and this is the approach used throughout this book.

As shown later, in addition to their use in determining the voltage drop of a circuit, they can also be used, equally directly, to determine the resistance per metre of a circuit conductor. The resistance per metre (in milliohms/m) of a particular conductor in a single-phase or d.c. circuit is simply the tabulated mV/A/m value divided by 2 or, in a three-phase circuit, its tabulated mV/A/m value divided by $\sqrt{3}$.

Example 2.1

A d.c. circuit is wired in single-core 70°C pvc-insulated non-sheathed cable to BS 6004 having copper conductors of 10 mm² cross-sectional area.

If $I_b = 40$ A and $1 = 33$ m what is the voltage drop?

Answer

From Table 4DIB Column 2 the mV/A/m is found to be 4.4 milliohms/m.
The voltage drop is:

$$\frac{4.4 \times 40 \times 33}{1000} \text{V} = 5.8 \text{V}$$

Example 2.2

A single-phase circuit is wired in 10 mm^2 two-core mineral-insulated cable to BS 6207 having copper conductors and sheath and an overall covering of pvc.

If $I_b = 65$ A and $l = 40$ m what is the voltage drop?

Answer

The appropriate table of voltage drop is Table 4J1B and from Column 2 it is found that the mV/A/m is 4.2 milliohms/m.

The voltage drop is:

$$\frac{4.2 \times 65 \times 40}{1000} \text{V} = 10.9 \text{V}$$

Example 2.3

A 400 V three-phase circuit is to be wired in a four-core armoured cable to BS 5467 having XLPE insulation and aluminium conductors of 35 mm^2 cross-sectional area.

If $I_b = 120$ A and $l = 27$ m what is the percentage voltage drop?

Answer

From Table 4L4B Column 4 the $(\text{mV/A/m})_z$ is found to be 1.95 milliohms/m.

The voltage drop is:

$$\frac{1.95 \times 120 \times 27}{1000} \text{V} = 6.32 \text{V}$$

The percentage voltage drop is:

$$\frac{6.32 \times 100}{400} \% = 1.58\%$$

It is frequently necessary to determine the voltage at a point of utilisation of a final circuit fed from a sub-distribution board and the next example shows one method that can be used. It will be seen that it is necessary to assume that the distribution cable feeding the sub-distribution board is carrying the design current determined from the application of a diversity factor.

Example 2.4

A 230 V single-phase a.c. circuit is wired in two-core 70°C pvc-insulated and sheathed non-armoured cable to BS 6004 having copper conductors.

The circuit length is 20 m and $I_b = 25$ A. The conductor cross-sectional area of this cable is 6 mm^2.

This circuit is taken from a sub-distribution board which is supplied by a 400 V three-phase and neutral circuit wired in multicore 70°C pvc-insulated armoured cable to BS 6346, having copper conductors. This cable is clipped direct to a wall. Its cross-sectional area is 25 mm^2 and taking into account diversity the distribution circuit design current is taken to be 100 A. The length of this circuit is 30 m.

What is the voltage at the point of utilisation of the single-phase final circuit?

Answer

First calculate the voltage drop in the distribution circuit.
From Table 4D4B Column 4 the $(mV/A/m)_z$ is found to be 1.5 milli-ohms/m.

Therefore the voltage drop is:

$$\frac{1.5 \times 100 \times 30}{1000} V = 4.5 V$$

Thus the line-to-line voltage is 400 – 4.5 V = 395.5 V.

The line-to-neutral voltage is:

$$\frac{395.5}{\sqrt{3}} V = 228.3 V$$

The voltage drop in the final circuit is then calculated as follows:

From Table 4D2B Column 3, the mV/A/m is found to be 7.3 milli-ohms/m.

The voltage drop is given by:

$$\frac{7.3 \times 20 \times 25}{1000} V = 3.65 V$$

The voltage at the point of utilisation is:

$$228.3 - 3.65 V = 224.7 V$$

It is accepted that in many practical cases some imbalance may occur on the three-phase side, leading to current in the neutral conductor of the distribution circuit. What the magnitude of this current could be is very difficult, and in many cases impossible, to forecast. For this reason alone it is necessary to assume, as has been done in Example 2.4, that balanced conditions apply.

THE MORE ACCURATE APPROACH TAKING ACCOUNT OF CONDUCTOR OPERATING TEMPERATURE

Earlier in this chapter, when introducing the simple approach, it was stated that four particular factors were not needed. The more accurate approach, however, does require the designer to know two of those four factors, namely:

(i) the ambient temperature
(ii) whether the circuit is to be run singly or grouped with other circuits.

Where the load power factor is known, as will be shown later, it is sometimes of advantage to take this into account.

The tabulated mV/A/m values for conductor cross-sectional areas of 16 mm^2 or less and the tabulated $(mV/A/m)_r$ values for 25 mm^2 or greater are based on the conductor actual operating temperature being the maximum permitted normal operating temperature for the type of cable insulation concerned, as indicated in the relevant table of current-carrying capacity.

It will be sheer coincidence if this equality between the two temperatures is obtained in practice and the most common situation is that the conductor operating temperature will be less than the maximum permitted because of compliance with Regulation 523–01–01.

It follows that if the designer so wishes he can use a lower value of mV/A/m or of $(mV/A/m)_r$ than the tabulated value for design purposes providing he is able to calculate first the actual operating temperature.

For circuits run singly and not totally embedded in thermally insulating material the conductor's actual operating temperature is given by:

$$t_l = t_a + \frac{I_b^2}{I_{ta}^2}(t_p - t_r)^\circ C$$

and

$$\frac{\text{design mV/A/m}}{\text{tabulated mV/A/m}} = \frac{230 + t_l}{230 + t_p}$$

The second equation is based on the approximate value of resistance–temperature coefficient of 0.004 per °C at 20°C which is applicable to both copper and aluminium conductors.

Example 2.5

A single-phase circuit is wired in two-core armoured 70°C pvc-insulated cable with 16 mm^2 copper conductors. The cable is clipped direct to a wall and is not grouped with other cables.

If $I_b = 70$ A, $t_a = 35$°C and $l = 30$ m what is the voltage drop?

Answer

From Table 4D4A Column 2, $t_p = 70$°C and $I_{ta} = 89$ A. Also $t_r = 30$°C.

Thus the conductor operating temperature is given by:

$$t_l = 35 + \frac{70^2}{89^2}(70 - 30)°\text{C} = 59.7°\text{C}$$

From Table 4D4B Column 3, the tabulated mV/A/m is 2.8 milliohms/m.

So the voltage drop is:

$$\left(\frac{230 + 59.7}{230 + 70}\right)\left(\frac{2.8 \times 70 \times 30}{1000}\right)\text{V} = 5.7\,\text{V}$$

Example 2.6

A three-phase circuit is run in multicore sheathed and non-armoured cable having 85°C rubber insulation and copper conductors of 10 mm^2 cross-sectional area. The cable is mounted in free air (Reference Method 13) and is not grouped with other cables.

If $t_a = 50°\text{C}$, $l = 20$ m and $I_b = 45\,\text{A}$ what is the voltage drop?

Answer

From Table 4F2A Column 7, $t_p = 85°\text{C}$ and $I_{ta} = 71\,\text{A}$. Also $t_r = 30°\text{C}$.

Thus:

$$t_l = 50 + \frac{45^2}{71^2}(85 - 30)°\text{C} = 72.1°\text{C}$$

From Table 4F2B Column 4, the tabulated mV/A/m is 4 milliohms/m.

The voltage drop is:

$$\left(\frac{230 + 72.1}{230 + 85}\right)\left(\frac{4 \times 45 \times 20}{1000}\right)\text{V} = 3.45\,\text{V}$$

When a circuit is protected by a semi-enclosed fuse to BS 3036 the conductor cross-sectional area one has to use is greater than if the protective device was an HBC fuse or miniature circuit breaker (of the same current rating). It follows that the ratio I_b/I_{ta} will be less as will then be the actual operating temperature.

In those cases where the BS 3036 fuse is providing overload protection and the reference ambient temperature is taken to be the expected value, it can be shown that the conductor operating temperature with 70°C pvc-insulated cables does not exceed 51°C, or, for 85°C rubber-insulated cables, 58.9°C.

Thus, when using semi-enclosed fuses to BS 3036, another option open to the designer is that instead of calculating the actual conductor operating temperature the following factors may be applied as multipliers to the tabulated mV/A/m values in order to obtain the design values, provided the expected ambient temperature is 30°C.

70°C pvc-insulated cables: 0.94
85°C rubber-insulated cables: 0.917

In some cases it will be found that by using this option there will be no need to increase the conductor cross-sectional area from voltage drop considerations.

Example 2.7

A single-phase circuit having $I_b = 18\,A$ is protected by a 20 A BS 3036 semi-enclosed fuse against overload and short circuit. The circuit is wired in $6\,mm^2$ flat 70°C pvc-insulated and sheathed cables to BS 6004 having copper conductors.

If $l = 48\,m$ and $t_a = 30°C$ what is the voltage drop?

Answer

From Table 4D2B Column 3 the mV/A/m value is 7.3 milliohms/m.

The *design* mV/A/m = 7.3 × 0.94 milliohms/m = 6.862 milliohms/m.

The voltage drop is:

$$\frac{18 \times 48 \times 6.862}{1000}\,V = 5.93\,V$$

Example 2.8

A three-phase circuit having $I_b = 40\,A$ is protected by 45 A BS 3036 semi-enclosed fuses against overload and short circuit. The circuit is wired in single-core non-armoured cables having 85°C rubber insulation and $10\,mm^2$ copper conductors in trefoil touching.

If $l = 65\,m$ and $t_a = 30°C$ what is the voltage drop?

Answer

From Table 4F1B Column 7 the mV/A/m value is 4 milliohms/m.

The *design* mV/A/m = 4 × 0.917 milliohms/m = 3.67 milliohms/m.

The voltage drop is:

$$\frac{40 \times 65 \times 3.67}{1000}\,V = 9.54\,V$$

Example 2.9

A single-phase circuit having $I_b = 26\,A$ is to be protected by a 30 A BS 3036 semi-enclosed fuse against overload and short circuit. The circuit is to be wired in two-core 70°C pvc-insulated and sheathed non-armoured cable embedded in plaster, not grouped with other cables.

If $l = 33\,m$ and $t_a = 35°C$ what is the minimum conductor cross-sectional area that can be used and what is the voltage drop?

Answer

From Table 42C, $C_a = 0.97$

As $C_g = 1$ and $C_i = 1$

Thus:

$$I_t = 30 \times \frac{1}{0.97} \times \frac{1}{0.725} \text{ A} = 42.66 \text{ A}$$

From Table 4D2A Column 6 it is found that the minimum conductor cross-sectional area that can be used is 6 mm^2 having $I_{ta} = 46$ A.

The conductor operating temperature is given by:

$$t_l = 35 + \frac{26^2}{46^2}(70 - 30)°\text{C} = 47.8°\text{C}$$

From Table 4D2B Column 3 the mV/A/m is found to be 7.3 milli-ohms/m.

The voltage drop is:

$$\left(\frac{230 + 47.8}{230 + 70}\right)\left(\frac{7.3 \times 26 \times 33}{1000}\right) \text{V} = 5.8 \text{ V}$$

In order to show very rapidly the sort of advantage one can gain by taking into account conductor operating temperature, Figure 2.1 gives a family of curves of the reduction factor, i.e. the ratio of the design mV/A/m to the tabulated mV/A/m plotted against actual ambient temperature, each curve being for a particular value of I_b/I_{ta}. Figure 2.1 is for 70°C pvc-insulated cables and the two other similar figures – Figures 2.2 and 2.3 – are for 85°C rubber-insulated cables and cables having XLPE insulation respectively. These reduction factors, for the larger cables, apply only to the $(\text{mV/A/m})_r$ values and not to the $(\text{mV/A/m})_x$ values.

The following two examples show how these figures may be used.

Example 2.10

A three-phase circuit having $I_b = 35$ A is to be protected against overload by 45 A BS 1361 fuses. This circuit is to be wired in a multicore 70°C pvc-insulated armoured cable having copper conductors to BS 6346, clipped direct and not grouped with other cables.

If $t_a = 30°$C and $l = 55$ m what is the minimum conductor cross-sectional area that can be used and what is the voltage drop?

Answer

As C_a, C_g and C_i do not apply

$$I_t = I_n = 45 \text{ A}$$

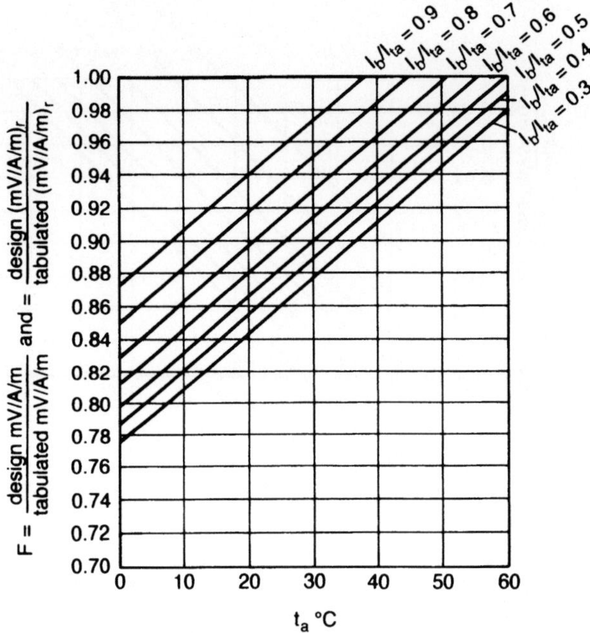

Figure 2.1 Reduction factors for different ambient temperatures and values of I_b/I_{ta} – 70°C pvc-insulated cables – circuits run singly.

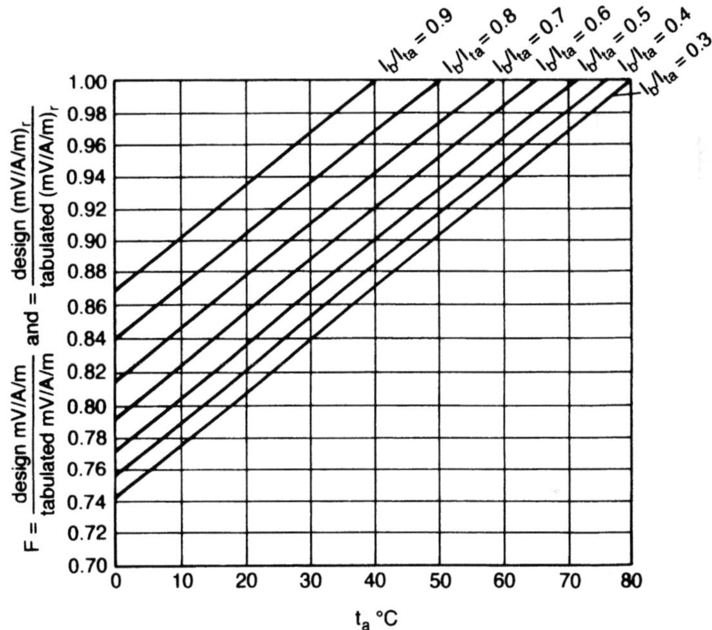

Figure 2.2 Reduction factors for different ambient temperatures and values of I_b/I_{ta} – 85°C rubber-insulated cables – circuits run singly.

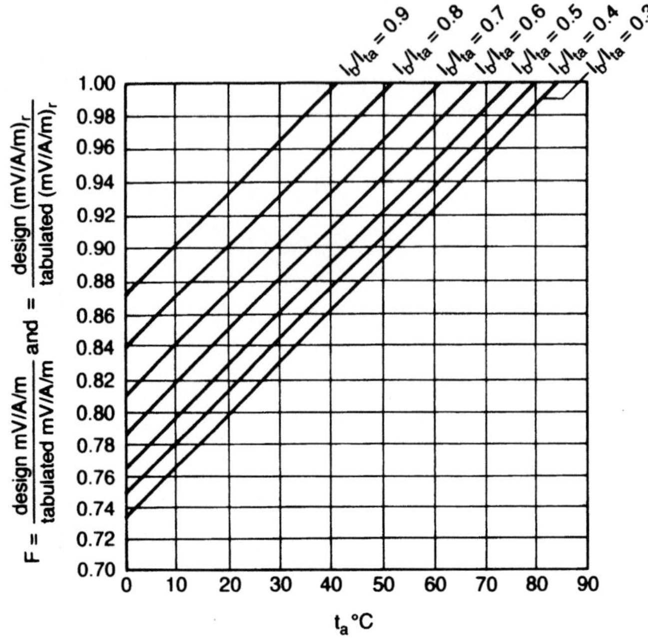

Figure 2.3 Reduction factors for different ambient temperatures and values of I_b/I_{ta} – cables having XLPE insulation – circuits run singly.

From Table 4D4A Column 3 it is found that the minimum conductor cross-sectional area that can be used is 10 mm^2 having $I_{ta} = 58$ A and tabulated mV/A/m of 3.8 milliohms/m (Column 4 in Table 4D4B).

Thus: $I_b/I_{ta} = 35/58 = 0.603$

From Figure 2.1 for this value of I_b/I_{ta} and $t_a = 30$°C it is found that the reduction factor (F) is 0.915.

The *design* mV/A/m is then 0.915×3.8 milliohms/m $= 3.477$ milliohms/m and the voltage drop is:

$$\frac{3.447 \times 35 \times 55}{1000} \text{ V} = 6.7 \text{ V}$$

Example 2.11

A three-phase circuit is wired in 50 mm^2 multicore armoured cable having XLPE insulation and aluminium conductors. It is run in free air (Reference Method 13) but is not grouped with other cables.

If $I_b = 110$ A, $t_a = 20$°C and $l = 35$ m what is the voltage drop? The power factor of the load is not known.

Answer

From Table 4L4B Column 4, $(mV/A/m)_r = 1.40$ milliohms/m and $(mV/A/m)_x = 0.135$ milliohms/m.

From Table 4L4A Column 5, $I_{ta} = 145\,A$.

Thus: $I_b/I_{ta} = 110/145 = 0.76$.

From Figure 2.3 the reduction factor F is found to be 0.89.

Hence *design* $(mV/A/m)_r = 0.89 \times 1.40$ milliohms/m $= 1.246$ milliohms/m.

'Corrected' $(mV/A/m)_z = \sqrt{1.246^2 + 0.135^2} = 1.253$ milliohms/m.

Voltage drop is:

$$\frac{1.253 \times 110 \times 35}{1000}\,V = 4.82\,V$$

So far in this chapter all the examples have concerned circuits run singly, i.e. not grouped or bunched with other circuits. For grouped circuits the *simple* approach for the calculation of voltage drop is exactly the same as for single circuits and the same is true of the method of taking account of the load power factor.

Example 2.12

Six similar single-phase circuits are installed in a common trunking. Each has $I_b = 27\,A$ and is protected against both overload and short circuit by a 30 A mcb. The circuits are run in single-core 70°C pvc-insulated cables to BS 6004.

If $t_a = 45°C$ and $l = 50\,m$ what is the minimum conductor cross-sectional area that can be used and what is the voltage drop?

Answer

From Table 4B1, $C_g = 0.57$.

From Table 4C1, $C_a = 0.79$ so that:

$$I_t = 30 \times \frac{1}{0.57} \times \frac{1}{0.79}\,A = 66.6\,A$$

From Table 4D1A, Column 4 it will be found that the minimum conductor cross-sectional area that can be used is $16\,mm^2$ having $I_{ta} = 76\,A$.

From Table 4D1B, Column 3 the mV/A/m value for this conductor is 2.8 milliohms/m.

The voltage drop is given by:

$$\frac{2.8 \times 27 \times 50}{1000}\,V = 3.78\,V$$

In this particular example, had the circuits been 230 V circuits subject to a maximum voltage drop of 2.5%, i.e. of 5.8 V, there would have been no need to use other than this simple approach to calculate the voltage drop.

Example 2.13

Eight similar single-phase circuits are grouped together, clipped direct, each having $I_b = 24\,A$ and protected against overload and short circuit by a 30 A BS 3036 semi-enclosed fuse. The circuits are run in non-armoured 70°C two-core (with cpc) pvc-insulated and sheathed cables to BS 6004.

If $t_a = 30°C$ and $l = 70\,m$ what is the minimum conductor cross-sectional area that can be used and what is the voltage drop?

Answer

From Table 4B1, $C_g = 0.52$

$C_a = 1$ and $C_d = 0.725$

$$I_t = 30 \times \frac{1}{0.52} \times \frac{1}{0.725}\,A = 79.6\,A$$

From Table 4D5A, Column 4 it is found that the minimum conductor cross-sectional area that can be used is $16\,mm^2$ having $I_{ta} = 85\,A$.

From Table 4D2B, Column 3 the mV/A/m value is found to be 2.8 milliohms/m.

The voltage drop is given by:

$$\frac{2.8 \times 24 \times 70}{1000}\,V = 4.7\,V$$

The next example concerns cables in parallel.

Example 2.14

A three-phase circuit comprises two multicore non-armoured 70°C pvc-insulated cables having copper conductors of $185\,mm^2$ cross-sectional area.

If $I_b = 550\,A$ and $l = 60\,m$ what is the voltage drop?

Answer

From Table 4D2B, Column 4 the $(mV/A/m)_z$ value is found to be 0.25 milliohms/m.

The voltage drop line-to-line is given by:

$$\frac{550 \times 60 \times 0.25}{2 \times 1000}\,A = 4.13\,V$$

Note the introduction of the factor '2' in the denominator, because there are 2 cables in parallel.

Now consider the more accurate approach for grouped circuits.

When the overcurrent protective devices are BS 3036 semi-enclosed fuses the most practical way of obtaining a more accurate calculated value of the voltage drop is to use the factor of 0.94. This accounts for the fact that the conductor temperature does not exceed 51°C if the cables are 70°C pvc-insulated or the factor 0.917 if the cables are 85°C rubber-insulated.

When the overcurrent protection is provided by HBC fuses or mcbs *and* the expected ambient temperature is 30°C or greater, the actual conductor temperature is given by:

$$t_l = t_p - \left(C_a^2 C_g^2 - \frac{I_b^2}{I_{ta}^2}\right)(t_p - 30)°C$$

and

$$\frac{\text{design mV/A/m}}{\text{tabulated mV/A/m}} = \frac{230 + t_l}{230 + t_p}$$

For conductor cross-sectional areas greater than 16 mm^2 the above relationship applies only to the resistance component of the voltage drop, i.e.

$$\frac{\text{design (mV/A/m)}_r}{\text{tabulated (mV/A/m)}_r} = \frac{230 + t_l}{230 + t_p}$$

but the design $(mV/A/m)_x$ = tabulated $(mV/A/m)_x$ because reactance is not influenced by temperature.

Example 2.15

Five similar single-phase circuits are grouped together, clipped direct, and run in two-core non-armoured 70°C pvc-insulated and sheathed cables to BS 6004 having copper conductors of 16 mm^2 cross-sectional area. Each circuit has I_b = 29 A and is protected against short circuit currents only by a BS 88 'gG' fuse.

If t_a = 50°C and l = 70 m what is the voltage drop?

Answer

From Table 4B1, $C_g = 0.60$

From Table 4C1, $C_a = 0.71$ and $t_p = 70°C$

From Table 4D2A Column 6, $I_{ta} = 85$ A

$$t_l = 70 - \left(0.71^2 \times 0.60^2 - \frac{29^2}{85^2}\right)(70 - 30)°C$$

$$= 70 - 2.6°C$$

$$= 67.4°C$$

From Table 4D2B Column 3, mV/A/m = 2.8 milliohms/m.

$$\text{Design mV/A/m} = \left(\frac{230 + 67.4}{230 + 70}\right) \times 2.8 = 2.78\,\text{milliohms/m}$$

$$\text{Voltage drop} = \frac{2.78 \times 29 \times 70}{1000}\,\text{V} = 5.64\,\text{V}$$

In this particular example the calculated conductor operating temperature is only marginally less than the maximum permitted normal operating temperature and there has been no real reduction gained in the calculated voltage drop. The problem is that from examination of the formula used to determine t_1 it is very difficult to forecast what reduction will be obtained in the voltage drop.

In order to show the reduction in mV/A/m or $(\text{mV/A/m})_r$ that is obtained for grouped circuits, Figures 2.4, 2.5 and 2.6 have been prepared for 70°C pvc-insulated cables, 85°C rubber-insulated cables and cables having XLPE insulation respectively.

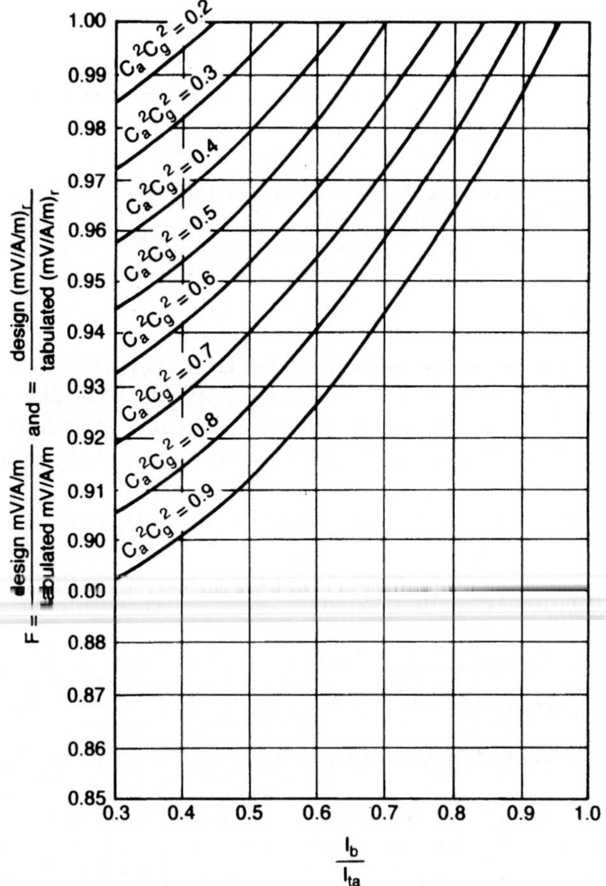

Figure 2.4 Reduction factors for 70°C pvc-insulated cables, grouped, in ambient temperature of 30°C or greater.

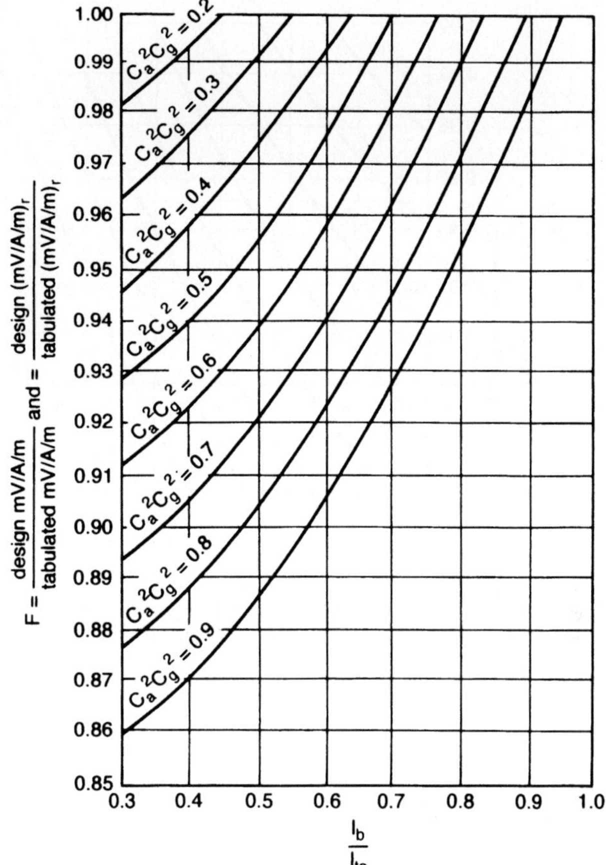

Figure 2.5 Reduction factors for 85°C rubber-insulated cables, grouped, in ambient temperature of 30°C or greater.

Each of these gives a family of curves of the reduction factor F plotted against I_b/I_{ta}, each curve in the family being for a particular value of $C_a^2C_g^2$.

The use of these families of curves enables the designer to reasonably quickly establish the extent of the advantage found in any particular case, if the more accurate method of calculating voltage drop is used.

In some cases the reduction in the mV/A/m or $(\text{mV/A/m})_r$ value that the more accurate method allows the designer to use may show that it is not necessary to increase the conductor cross-sectional area.

In Chapter 1 mention has already been made of the fact that when the expected ambient temperature is less than 30°C and where the overcurrent protective device is providing overload protection, with or without short circuit protection, the limiting aspect when determining the ambient temperature correction factor is the need not to exceed the maximum tolerable conductor temperature under overload conditions. This, in turn, means that under normal load conditions the conductors will be operating at less than maximum normal operating temperature.

For circuits operating at ambient temperatures below 30°C and protected by other than BS 3036 semi-enclosed fuses against both overload and short circuit the

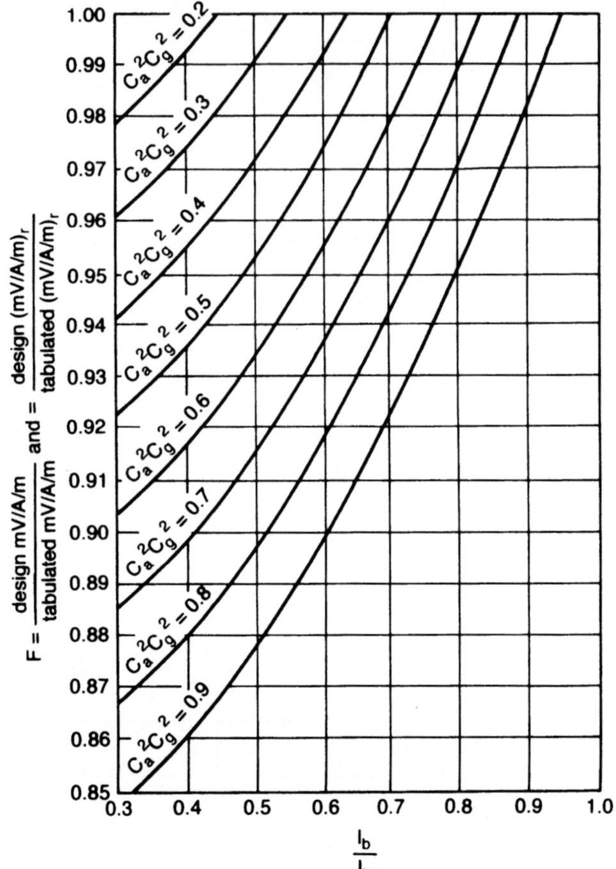

Figure 2.6 Reduction factors for cables having XLPE insulation, grouped, in ambient temperature of 30°C or greater.

following reduction factors may be applied to the tabulated mV/A/m or $(mV/A/m)_r$ values in order to obtain the design values.

Ambient temperature	Reduction factor
0°C	0.96
10°C	0.97
20°C	0.985

The factors apply to both copper and aluminium conductors and pvc-insulation (both 70°C and 85°C types), 85°C rubber insulation and XLPE insulation.

Little real advantage is found from using these reduction factors, particularly for the larger cable sizes, because they are applied only to the resistive component i.e. to the $(mV/A/m)_r$ values. It will be seen from the volt drop tables in Appendix 4 of BS 7671 for the largest conductor sizes, particularly for single-core non-armoured cables, that the $(mV/A/m)_r$ value is very small compared with the $(mV/A/m)_x$ value. For such cables there is no point in taking account of conductor temperature.

THE MORE ACCURATE APPROACH TAKING ACCOUNT OF LOAD POWER FACTOR

The other factor which influences voltage drop is that of the power factor of the load.

Where the power factor of the load is not known the designer has no option but to use the equations given at the commencement of this chapter for the simple approach, with or without adjustment for the conductor operating temperature.

Where the power factor, $\cos\phi$, of the load *is* known it is worthwhile using the following formulae:

For a.c. circuits using conductors of $16\,\text{mm}^2$ or less cross-sectional area:

$$\text{Voltage drop} = \frac{\text{tabulated mV/A/m} \times I_b \times l \times \cos\phi}{1000}\ \text{V}$$

For a.c. circuits using conductors of $25\,\text{mm}^2$ or greater cross-sectional area, the voltage drop is given by:

$$\frac{[\text{tabulated (mV/A/m)}_r \times \cos\phi + \text{tabulated (mV/A/m)}_x \times \sin\phi] \times I_b \times l}{1000}\ \text{V}$$

where $\sin\phi = \sqrt{1 - \cos^2\phi}$.

Example 2.16

A single-phase circuit having $I_b = 27\,\text{A}$ is run in single-core 70°C pvc-insulated, non-sheathed, non-armoured cables having copper conductors and is enclosed in conduit on a wall with five other similar circuits. It is protected against both overload and short circuit by means of a 30 A mcb.

If $t_a = 40°\text{C}$, $l = 38\,\text{m}$ and the power factor of the load is 0.8 lagging, what is the minimum conductor cross-sectional area that can be used and what is the voltage drop?

Answer

From Table 4B1, $C_g = 0.57$

From Table 4C1, $C_a = 0.87$

As $C_i = 1$

$$I_t = 30 \times \frac{1}{0.57} \times \frac{1}{0.87}\ \text{A} = 60.5\ \text{A}$$

From Table 4D1A, Column 4 it is found that the minimum conductor cross-sectional area that can be used is $16\,\text{mm}^2$ having $I_{ta} = 76\,\text{A}$.

From Table 4D1B Column 3, the mV/A/m value is 2.8 milliohms/m.

$$\text{Voltage drop} = \frac{2.8 \times 27 \times 38 \times 0.8}{1000}\ \text{V} = 2.3\ \text{V}$$

In this example no attempt has been made to take account of the fact that the conductor operating temperature is less than the maximum permitted value.

Example 2.17

A single-phase circuit is wired in two-core armoured cable having XLPE insulation to BS 5467 and 25 mm^2 copper conductors.

If $I_b = 120$ A and $l = 30$ m what is the voltage drop if the power factor of the load is 0.7 lagging?

Answer

From Table 4E4B Column 3, $(mV/A/m)_r = 1.85$ milliohms/m and $(mV/A/m)_x = 0.16$ milliohms/m.

With $\cos\phi = 0.7$, $\sin\phi = \sqrt{1 - 0.7^2} = 0.714$

The voltage drop is given by:

$$\frac{(1.85 \times 0.7 + 0.16 \times 0.714) \times 120 \times 30}{1000} \text{ V} = 5.07 \text{ V}$$

Had the power factor been ignored the voltage drop would have been calculated as:

$$\frac{1.9 \times 120 \times 30}{1000} \text{ V} = 6.84 \text{ V}$$

If then the circuit had a nominal voltage of 230 volts and been limited to a 2.5% voltage drop (i.e. 5.8 volts) the designer would have had apparently to use the larger conductor cross-sectional area of 35 mm^2.

Figure 2.7 has been developed to show the advantage that may be gained by basing the voltage drop calculation on the more accurate $[(mV/A/m)_r \cos\phi + (mV/A/m)_x \sin\phi]$ as compared with basing it on $(mV/A/m)_z$.

For example a three-phase circuit is run in single-core armoured 70°C pvc-insulated cables having 400 mm^2 copper conductors, in trefoil, touching. The power factor of the load is 0.9 lagging. From Table 4D3B it is found that $(mV/A/m)_r = 0.12$ milliohms/m and $(mV/A/m)_z = 0.195$ milliohms/m.

Thus:

$$\frac{(mV/A/m)_r}{(mV/A/m)_z} = \frac{0.12}{0.195} = 0.615$$

Figure 2.7 then indicates that calculating the voltage drop using $[(mV/A/m)_r \cos\phi + (mV/A/m)_x \sin\phi]$ would give a value 0.9 times that calculated using $(mV/A/m)_z$.

As already stated, examination of the volt drop sections of the tables in Appendix 4 of BS 7671 indicates that for conductors of the largest cross-sectional areas the $(mV/A/m)_r$ values are much less than the $(mV/A/m)_x$ values. For such cases one can ignore the $(mV/A/m)_r$ values and calculate the voltage drop from:

$$\frac{(mV/A/m)_x \times \sin\phi \times I_b \times l}{1000} \text{ V}$$

or, of course, the designer may prefer to use the $(mV/A/m)_z$ approach.

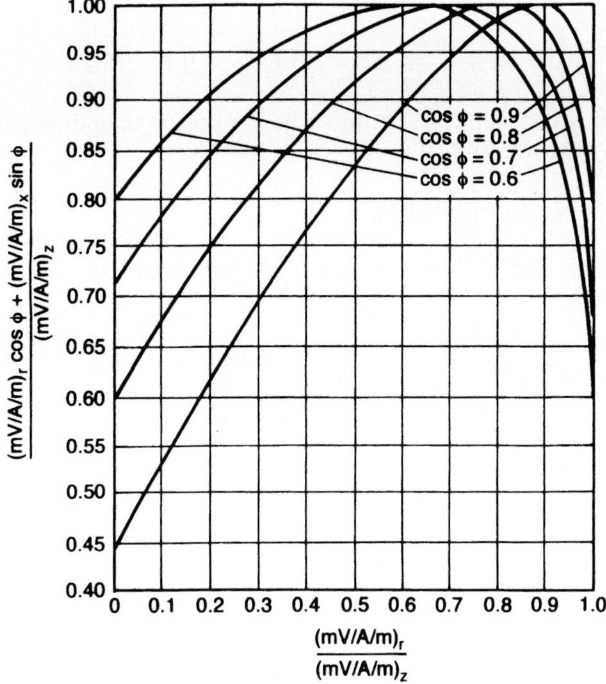

Figure 2.7 Showing reduction in calculated voltage drop if load power factor is taken into account.

THE MORE ACCURATE APPROACH TAKING ACCOUNT OF BOTH CONDUCTOR OPERATING TEMPERATURE AND LOAD POWER FACTOR

The following examples show the calculations that are necessary where the designer wishes to take account of both the conductor operating temperature *and* the load power factor.

Example 2.18

A single-phase circuit, run singly, uses two-core 85°C rubber-insulated cable, clipped direct. The conductors are of copper and their cross-sectional area is 16 mm^2.

If $I_b = 65\,A$, $t_a = 50°C$, the load power factor is 0.7 lagging and $l = 20\,m$, what is the voltage drop?

Answer

From Table 4F2A, Column 4, $I_{ta} = 103\,A$

$t_p = 85°C$ and $t_r = 30°C$.

$$t_l = 50 + \frac{65^2}{103^2}(85 - 30)°C = 71.9°C$$

Thus the reduction factor to be applied to the tabulated mV/A/m value to take account of t_l is

$$\frac{230 + 71.9}{230 + 85} = 0.96$$

From Table 4F2B Column 3 the tabulated mV/A/m value is 2.9 milliohms/m.

The voltage drop is therefore given by:

$$\frac{2.9 \times 0.96 \times 0.7 \times 65 \times 20}{1000} V = 2.5 V$$

It will be noted that the reduction in the voltage drop caused by taking account of the load power factor is far greater than that caused by taking account of the conductor operating temperature.

Example 2.19

A three-phase circuit is run, clipped direct and not grouped with other circuits, in a multicore armoured 70°C pvc-insulated cable having 95 mm^2 copper conductors.

If $I_b = 160$ A, $t_a = 45°C$, the load power is 0.8 lagging and $l = 120$ m, what is the voltage drop?

Answer

From Table 4D4A Column 3, $I_{ta} = 231$ A, and from Column 4 of Table 4D4B $(mV/A/m)_r = 0.41$ milliohms/m and $(mV/A/m)_x = 0.135$ milliohms/m.

$t_p = 70°C$ and $t_r = 30°C$.

Thus:

$$t_l = 45 + \frac{160^2}{231^2}(70 - 30)°C = 64.2°C$$

The design $(mV/A/m)_r$ taking account of conductor operating temperature is given by:

$$\left(\frac{230 + 64.2}{230 + 70}\right) \times 0.41 \text{ milliohms/m} = 0.402 \text{ milliohms/m}$$

The design $(mV/A/m)_x$ remains the same as the tabulated value.

The voltage drop is given by:

$$[0.8 \times 0.402 + \sqrt{(1 - 0.8^2)} \times 0.135] \times \frac{160 \times 120}{1000} V = 7.73 V$$

VOLTAGE DROP IN RING CIRCUITS

Occasions may arise when it is necessary to calculate the voltage drop occurring in a ring circuit. The method to use is illustrated by Example 2.20 if the loads taken from the points of utilisation and the cable lengths between those points are known.

Example 2.20

Figure 2.8(a) shows a single-phase ring circuit wired in $2.5\,mm^2$ flat two core (with cpc) 70°C pvc-insulated and sheathed cable to BS 6004. The figure gives the loads taken from each point of utilisation.

The first stage is to determine the current distribution and as shown in Figure 2.8(b) a current given by I_x A is taken to flow in the first section, $(I_x - 5)$A in the second section, and so on.

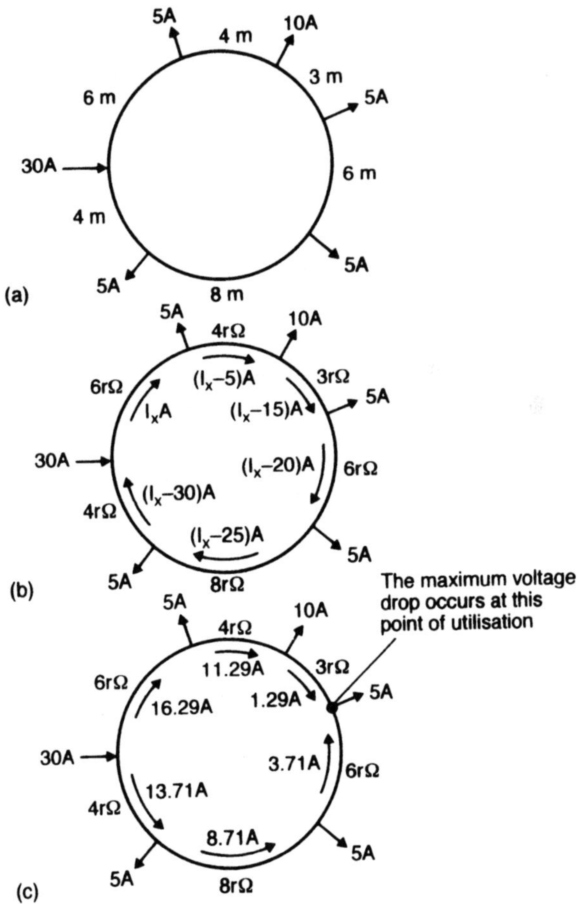

Figure 2.8 Voltage drop in ring circuit.

If the resistance per metre of the phase conductor is denoted by 'r' then:

$$I_x 6r + (I_x - 5)4r + (I_x - 15)3r + (I_x - 20)6r + (I_x - 25)8r + (I_x - 30)4r = 0$$

From which:

$$I_x r(6 + 4 + 3 + 6 + 8 + 4) - r(20 + 45 + 120 + 200 + 120) = 0$$

Thus: $I_x = 16.29\,A$

The current distribution therefore is as shown in Figure 2.8(c) and it is now possible to calculate the voltage drop.

This is given by:

$$[(16.29 \times 6) + (11.29 \times 4) + (1.29 \times 3)] \times \frac{mV/A/m}{1000}\,V$$

i.e.

$$\frac{146.8 \times mV/A/m}{1000}\,V$$

From Table 4D2B Column 3 the mV/A/m value is found to be 18 milli-ohms/m.

The voltage drop is therefore

$$\frac{146.8 \times 18}{1000}\,V = 2.64\,V$$

This can be checked by calculating the voltage drop in the anti-clockwise direction to the point of utilisation at which the maximum value occurs.

It can be shown that the maximum voltage drop that can occur in a ring circuit is when a load current equal to the nominal current of the overcurrent protective device of the circuit is taken from the mid-point of that circuit. On this basis the maximum length from the origin of the circuit to its mid-point for a 230 V single-phase 30 A ring circuit having a permitted 2.5% voltage drop (i.e. 5.8 V) is given by:

$$\frac{6000}{15 \times 18}\,m = 22.2\,m$$

or, in other words, a maximum total length of 44.4 m.

Taking into account the fact that the circuit conductors are carrying only 15 A while their current-carrying capacity is 23 A if enclosed in conduit or 27 A if clipped direct (see Table 4D2A Column 4 and Table 4D5A Column 4) that maximum length of 44.4 m becomes 48 m or 49 m respectively.

However, if one assumes that the total load current, although still equal to the nominal current of the overcurrent protective device, is uniformly distributed around the ring, the maximum permitted length can be considerably greater.

For all the examples given in this chapter the design current of the circuit concerned has been stated and no indication has been given of the maximum permissible voltage drop.

The key regulation is Regulation 525–01–01 which does not directly specify what the latter shall be but requires the voltage at the terminals of any current-using equipment to be not less than the lower limit corresponding to the relevant British Standard. If that equipment is not the subject of a British Standard then the voltage at the terminals shall be such as not to impair the safe functioning of the equipment.

Regulation 525–01–02 states that if the voltage drop between the origin of the installation and the current-using equipment does not exceed 4% of the nominal voltage of the supply this is deemed to satisfy Regulation 525–01–01. Thus the designer could exceed 4% but still claim to satisfy Regulation 525–01–01. In determining what percentage could be allowed it must be borne in mind that the electricity supply industry has a statutory requirement to maintain the voltage at a consumer's terminals to within 94% to 110% of the nominal value.

It follows that having calculated the voltage drop inside the installation the designer should add 6% of the nominal voltage to that voltage drop in order to determine the voltage at the points of utilisation and it is this value which may have to be compared with the limits given in the relevant British Standard. It is felt there is no need to give an example illustrating this aspect but to devote the remainder of this chapter to the consideration of extra-low voltage circuits.

VOLTAGE DROP IN ELV CIRCUITS

Whichever type of extra-low voltage circuit is used, whether it be SELV, or FELV, the calculation of voltage drop is of importance, particularly when the nominal voltage of the circuit is low, e.g. 12 V. Many of these circuits are fed from transformers and the first essential information required by the designer is the regulation of the transformer, where:

$$\text{Regulation} = \frac{U_{nl} - U_{fl}}{U_{nl}} \times 100\%$$

where: U_{nl} = output voltage at no-load, in volts
U_{fl} = output voltage at full-load, in volts.

A 12 V luminaire may be fed from its own transformer in which case the output voltage at full load can be 12 V. If, however, a transformer is supplying a number of 12 V luminaires the full load output voltage will be less (say 11.8 V) in order to prevent overvoltage occurring should one of the lamps fail. Whichever method is used, the voltage at the luminaire should not be less than 11.4 V.

Example 2.21

A 50 watt 12 V luminaire is to be fed from a transformer having a full load output voltage of 12 V. If it is intended to use 70°C pvc-insulated and sheathed cable having 1.5 mm^2 copper conductors clipped direct, what is the maximum circuit length that can be tolerated if the voltage at the luminaire must not be less than 11.4 V? The ambient temperature $t_a = 30°C$.

Answer

From Table 4D2B, Column 3, the mV/A/m = 29 milliohms/m.

$$I_b = \frac{50}{12} A = 4.17\,A$$

Maximum permissible voltage drop = 0.6 V

Maximum permissible circuit length is given by:

$$\frac{0.6 \times 1000}{4.17 \times 29}\,m = 4.96\,m$$

It will be noted that I_{ta} = 19.5 A (from Column 6 of Table 4D2A) and it is of interest to establish the maximum circuit length using the more accurate approach.

$$t_l = 30 + \frac{4.17^2}{19.5^2}(70 - 30)°C = 31.8°C$$

The design mV/A/m is given by:

$$\left(\frac{230 + 31.8}{230 + 70}\right) \times 29\,\text{milliohms/m} = 25.3\,\text{milliohms/m}$$

The maximum circuit length is then given by:

$$\frac{0.6 \times 1000}{4.17 \times 25.3}\,m = 5.69\,m$$

Example 2.22

A 75 W 12 V luminaire together with others is to be fed from a transformer having a full load output voltage of 11.8 V. It is intended to use 70°C pvc-insulated cables having 2.5 mm^2 copper conductors in conduit.

If t_a = 30°C what is the maximum circuit length that can be tolerated if the voltage at the luminaire must not be less than 11.4 V.

Answer

From Table 4D2B, Column 3, mV/A/m = 18 milliohms/m.

$$I_b = \frac{75}{12} A = 6.25\,A$$

The maximum permissible voltage drop in the circuit is:

$$(11.8 - 11.4)\,V = 0.4\,V$$

Maximum permissible circuit length is given by:

$$\frac{0.4 \times 1000}{18 \times 6.25}\,m = 3.56\,m$$

Using the more accurate method, from Table 4D2A Column 4, I_{ta} = 23 A and

$$t_1 = 30 + \frac{6.25^2}{23^2}(70 - 30)°C = 33°C$$

The design mV/A/m is given by:

$$\left(\frac{230 + 33}{230 + 70}\right) \times 18 \text{ milliohms/m} = 15.8 \text{ milliohms/m}$$

The maximum circuit length is then given by:

$$\frac{0.4 \times 1000}{15.8 \times 6.25} \text{ m} = 4.05 \text{ m}$$

However, it can be argued that the more accurate method should not be used because it assumes only one luminaire is actually energised.

Chapter 3

Calculation of Earth Fault Loop Impedance

The most commonly used protective measure against indirect contact is that of automatic disconnection of supply using either the overcurrent protective devices, which are also providing protection against overload and/or short circuit currents for the circuits concerned, or residual current devices (rcds).

Whichever type of device is chosen it is necessary for the designer to calculate the earth fault loop impedance (Z_s) of every circuit in the installation in order to check that these impedances do not exceed the maxima specified in BS 7671.

Amendment 1 to BS 7671 recognised that the nominal supply voltage has been reduced from 240/415 V to 230/400 V. However this change is not reflected in the impedance values given in the tables of Chapter 41 which are now based on an assumed open circuit voltage of 240 V. In the examples given in the following chapters the nominal voltage, 230/400 V, has been used in the calculation of fault current. This assumption leads to prospective fault currents which are about 4% lower than would be calculated if an open circuit voltage of 240 V was assumed. This in turn leads to disconnection times which may be pessimistic, but safe. If the calculated fault currents are within 4% of the maximum breaking capacity of the protective device then it is recommended that the open circuit voltage, assumed to be 240 V, is used in the calculations.

Another reason for calculating the earth fault loop impedances is to check that the circuit protective conductors are adequately protected thermally, i.e. that they comply with the adiabatic equation given in Regulation 543–01–03. This aspect is the subject of Chapter 4.

For some circuits which normally are required to disconnect within 0.4 s, the designer has the option of increasing the maximum disconnection time to 5 s provided that the impedances of the circuit protective conductors do not exceed certain specified maxima (see Table 41C of BS 7671). When the designer uses this option it is necessary to calculate those impedances separately from the earth fault loop impedances.

One series of examples in this chapter shows the calculations that are necessary when the circuit conductors have a cross-sectional area of 35 mm^2 or less, where it has long been accepted that one can use conductor resistances instead of conductor impedances in those calculations. The other series of examples deals with the cases where conductor impedances should be used.

Regulation 413–02–05 requires that account is taken of the increase in temperature and resistance of conductors as a result of an overcurrent. Prior to Amendment 1 to BS 7671 it was usual for the protective conductor impedance to be based on the average temperature attained during the fault, that is the average of the assumed initial temperature and the maximum permitted final temperature. Similarly for the phase conductor impedance it was usual to use the average of the maximum permitted normal operating temperature and the maximum permitted final temperature.

Changes to Chapter 41 introduced in Amendment 1 allow a simpler approach to be taken under some conditions. For the devices listed in Appendix 3 of BS 7671, see below, the requirement of Regulation 413–02–05 is deemed to be satisfied if the circuit loop impedance meets the requirements of Regulations 413–02–10, 413–02–11 or 413–02–14 when the conductors are at their normal operating temperature.

(1) Fuses to BS 1361, up to 100 A
(2) Semi-enclosed fuses to BS 3036, up to 100 A
(3) Fuses to BS 88 Part 2 and Part 6, up to 200 A
(4) Types B, C and D mcbs to BS EN 60898 and RCBOs to BS EN 61009, up to 125 A

For these devices the normal operating temperature for the phase conductor is generally taken to be the maximum permitted normal operating temperature.

Selection of the normal operating temperature of the protective conductor is less obvious. For a protective conductor incorporated in a cable or bunched with other cables, Table 54C, the normal operating temperature is taken to be the maximum permitted normal operating temperature. For a protective conductor not incorporated in a cable and not bunched with other cables, Table 54B, it would appear reasonable to take the ambient temperature, 30°C, as the normal operating temperature. This approach is considered acceptable by most guidance on the subject. If the protective conductor has a much smaller cross-sectional area then its temperature rise under fault conditions will be greater than that of the phase conductor. In such a case it would be prudent to base the impedance of the protective conductor on the maximum permitted normal operating temperature.

This method is not rigorously correct because it assumes that the conductor impedance does not increase significantly during a fault. The method is intended to take account of the effect of normal load currents on the operating characteristics of some protective devices. The operating characteristics given in Appendix 3 of BS 7671 are based on the device being at ambient temperature at the start of the fault. When a device such as a fuse carries the normal load current its temperature will increase and its operating time, in the event of a fault, will be less than that given by the characteristics given in Appendix 3 of BS 7671. The same principles apply to a circuit breaker operating on the thermal part of its characteristic.

The assumptions made above regarding conductor temperatures during fault conditions should only be used when calculating circuit impedance. The values of k given in Chapter 54 of BS 7671 should not be adjusted for lower final temperatures as this will reduce the value of k and consequently increase the conductor size required to comply with Regulation 543–01–03.

If the actual conductor operating temperature is known to be less than the maximum permitted normal operating temperature then the actual temperature can be used when calculating circuit impedance. The actual conductor temperature can also be used as the initial temperature to calculate a revised value of k, see Chapter 4. This will be the case for circuits protected against overload by BS 3036 semi-enclosed fuses where the actual operating temperature will be less than the maximum potential operating temperature. See Chapter 2 and Example 3.8.

For devices other than those listed above it is necessary to calculate the earth fault loop impedance for the conductors based on the average of the initial temperature and the maximum permitted final temperature. Again, if the actual conductor operating temperature, or ambient temperature, is known these can be used to calculate the average temperature during the fault.

Consider first of all final circuits connected directly to the source at an installation's main distribution board (in the smaller installation, a consumer unit) and where the source is the public low voltage supply network.

The Electricity Council has stated, in Engineering Recommendation P23, that typical maximum values for Z_E, defined as that part of the earth fault loop impedance external to the installation, are 0.35 ohm for TN–C–S systems and 0.80 ohm for TN–S systems. These values are used for many of the examples for the purpose of illustration but in practice it is, of course, acceptable to use *measured* values of Z_E.

For the moment consideration is limited to cables having conductors of 35 mm² or less cross-sectional area and for the purposes of calculating Z_s their reactances can be ignored.

Thus for radial circuits:

$$Z_s = Z_E + R_1 + R_2 \text{ ohm}$$

and for ring circuits:

$$Z_s = Z_E + 0.25\,(R_{T1} + R_{T2})\text{ ohm}$$

where: R_{T1} = resistance of the phase conductor prior to connecting the ends to form a ring

R_{T2} = resistance of the protective conductor prior to connecting the ends to form a ring.

It is important to note that when Table 54B applies, the values shown in Columns 3, 5 and 7 of Table 3.1 can only be used for the combinations of phase conductor and protective conductors given in Columns 1 and 2 (when these conductors are of copper). Where the intended combination is not given, use the data given in Tables 3.3, 3.4 and 3.5.

For example, if the phase conductor has a cross-sectional area of 6 mm² and the protective conductor is 1.5 mm², $(R_1 + R_2)$/m for 85°C rubber insulation is *not* (3.88 + 15.2) milliohms/m, i.e. 19.08 milliohms/m. For the correct figure see Example 6.3.

However, when Table 54C applies, the values shown in Columns 4, 6 and 8 of Table 3.1 apply to any combinations of phase conductor and protective conductor, when they are of copper. A similar restriction applies to Table 3.2 when the conductors are of aluminium.

To assist the reader, Tables 3.1 and 3.2 have been developed for copper conductors and aluminium conductors respectively giving values for R_1/m and $(R_1 + R_2)$/m in milliohms per metre for 70°C pvc-insulated cable, 85°C rubber-insulated cable, and cables having XLPE insulation. The tabulated values have been based on the simplified formula given in BS 6360 which uses the resistance–temperature coefficient of 0.004 per °C at 20°C.

The values given in Tables 3.1 and 3.2 are based on assumed normal operating temperatures. When the protective conductor complies with the conditions of Table 54B its normal operating temperature is taken to be 30°C and that of the associated phase conductor is taken to be its maximum permitted normal operating temperature. When the protective conductor complies with the conditions

Table 3.1 Values of R_1/m and $(R_1 + R_2)$/m in milliohms/metre for copper conductors at their normal operating temperature.

Conductor cross-sectional area mm^2		70°C pvc insulation		85°C rubber insulation		90°C XLPE or pvc insulation	
Phase conductor	Protective conductor	When Table 54B applies	When Table 54C applies	When Table 54B applies	When Table 54C applies	When Table 54B applies	When Table 54C applies
1	2	3	4	5	6	7	8
1	—	21.7	21.7	22.8	22.8	23.2	23.2
1	1	40.5	43.4	41.6	45.6	42.0	46.3
1.5	—	14.5	14.5	15.2	15.2	15.5	15.5
1.5	1	33.3	36.2	34.1	38.1	34.3	38.7
1.5	1.5	27.1	29.0	27.8	30.5	28.1	31.0
2.5	—	8.89	8.89	9.34	9.34	9.48	9.48
2.5	1	27.7	30.6	28.2	32.1	28.3	32.7
2.5	1.5	21.5	23.4	21.9	24.6	22.1	25.0
2.5	2.5	16.6	17.8	17.0	18.7	17.2	19.0
4	—	5.53	5.53	5.81	5.81	5.90	5.90
4	1.5	18.1	20.1	18.4	21.1	18.5	21.4
4	2.5	13.2	14.4	13.5	15.1	13.6	15.4
4	4	10.3	11.1	10.6	11.6	10.7	11.8
6	—	3.70	3.70	3.88	3.88	3.94	3.94
6	2.5	11.4	12.6	11.6	13.2	11.6	13.4
6	4	8.49	9.23	8.68	9.69	8.74	9.84
6	6	6.90	7.39	7.08	7.76	7.15	7.88
10	—	2.20	2.20	2.31	2.31	2.34	2.34
10	4	6.99	7.73	7.10	8.11	7.14	8.24
10	6	5.40	5.89	5.51	6.19	5.55	6.28
10	10	4.10	4.39	4.21	4.61	4.25	4.68
16	—	1.38	1.38	1.45	1.45	1.47	1.47
16	6	4.58	5.08	4.65	5.33	4.68	5.41
16	10	3.28	3.58	3.35	3.75	3.38	3.81
16	16	2.58	2.76	2.65	2.90	2.67	2.94
25	—	0.872	0.872	0.916	0.916	0.931	0.931
25	10	2.78	3.07	2.82	3.22	2.83	3.27
25	16	2.07	2.25	2.11	2.37	2.13	2.40
25	25	1.63	1.74	1.67	1.83	1.69	1.86
35	—	0.629	0.629	0.660	0.660	0.671	0.671
35	16	1.82	2.01	1.86	2.11	1.87	2.14
35	25	1.38	1.50	1.42	1.58	1.43	1.60
35	35	1.17	1.26	1.21	1.32	1.22	1.34

of Table 54C the normal operating temperature of both the phase and protective conductors is taken to be the maximum permitted operating temperature.

When the protective device is not of a type listed above, values of R_1 and R_2 can be calculated from the values given in Table 3.3 by using the factors given in brackets in the lower half of Table 3.4.

Table 3.2 Values of R_1/m and $(R_1 + R_2)$/m in milliohms/metre for aluminium conductors at their normal operating temperature.

Conductor cross-sectional area mm^2		70°C pvc insulation		90°C XLPE or pvc insulation	
Phase conductor	Protective conductor	When Table 54B applies	When Table 54C applies	When Table 54B applies	When Table 54C applies
1	2	3	4	7	8
16	—	2.29	2.29	2.44	2.44
16	6	—	—	—	—
16	10	—	—	—	—
16	16	4.28	4.58	4.43	4.89
25	—	1.44	1.44	1.54	1.54
25	10	—	—	—	—
25	16	3.43	3.73	3.52	3.98
25	25	2.69	2.88	2.78	3.07
35	—	1.04	1.04	1.11	1.11
35	16	3.03	3.33	3.10	3.56
35	25	2.29	2.48	2.36	2.65
35	35	1.94	2.08	2.01	2.22

Table 52B of BS 7671 does not permit aluminium conductors of less than 16 mm^2 cross-sectional area.

THE SIMPLE APPROACH

Example 3.1

A single-phase circuit is run in single-core 70°C pvc-insulated and sheathed cables clipped direct and not bunched with cables of other circuits. Protection against indirect contact is provided by a device of a type listed earlier in this chapter.

If the conductors are copper, the cross-sectional area of the live conductors is 4 mm^2 and that of the protective conductor is 2.5 mm^2, l is 45 m and Z_E is assumed to be 0.35 ohm (the supply being PME), what is the earth fault loop impedance?

Answer

The relevant table is Table 54B so that the value of $(R_1 + R_2)$/m in milliohms/metre is obtained from Column 3 in Table 3.1. It is found to be 13.2 milliohms/m.

$$Z_s = 0.35 + \left(\frac{45 \times 13.2}{1000}\right) \text{ohm} = 0.94 \text{ ohm}$$

Note: As the protective conductor cross-sectional area does not comply with Table 54G it would be necessary to check that it was thermally protected by using the adiabatic equation of Regulation 543–01–03. In order to do this it is necessary to know the type and current rating of the protective device associated with the circuit and be in possession of its time/current characteristic. (See examples in Chapter 4.)

Example 3.2

A circuit is run in single-core 70°C pvc-insulated cables having copper conductors, these being bunched with cables of other circuits in conduit. Protection against indirect contact is provided by a device of a type listed earlier in this chapter.

If $l = 20\,m$, the cross-sectional area of the live conductors is $25\,mm^2$ and that of the protective conductor is $10\,mm^2$ and $Z_E = 0.8$ ohm, what is the earth fault loop impedance?

Answer

Because the circuit concerned is bunched with other circuits, the relevant table is Table 54C. $(R_1 + R_2)/m$ in milliohms per metre is obtained from Column 4 of Table 3.1 and is found to be 3.07 milliohms/m.

$$Z_s = 0.8 + \left(\frac{20 \times 3.07}{1000}\right) \text{ohm} = 0.861\ \text{ohm}$$

As in Example 3.1, the protective conductor does not comply with Table 54G and it would be necessary to check compliance with the adiabatic equation of Regulation 543–01–03.

Example 3.3

A single-phase ring circuit is run in $2.5\,mm^2$ 70°C pvc-insulated and sheathed flat cable to BS 6004 (the protective conductor having therefore a cross-sectional area of $1.5\ mm^2$). Protection against indirect contact is provided by a device of a type listed earlier in this chapter.

If $l = 65\,m$ and $Z_E = 0.35$ ohm, what is the earth fault loop impedance?

Answer

Because the protective conductor is an integral part of the cable, the relevant table is Table 54C and Column 4 of Table 3.1 applies.

From that column and table, $(R_1 + R_2)/m = 23.4$ milliohms/m

$$Z_s = 0.35 + \left(\frac{0.25 \times 65 \times 23.4}{1000}\right) \text{ohm} = 0.73\ \text{ohm}$$

Note the introduction of the factor 0.25 because the circuit concerned is a ring circuit. Also note the need to check against the adiabatic equation of Regulation 543–01–03.

In order to determine the earth fault loop impedance of a spur on a ring circuit, strictly it is necessary to estimate the fractional distance of that spur from the origin of the circuit. Denote that distance by 'y'.

Then the earth fault loop impedance at the remote end of the spur is given by:

$$Z_s = Z_E + y(1 - y)(R_{1T} + R_{2T}) + R_{1S} + R_{2S} \text{ ohm}$$

where: R_{1T} = total resistance of the phase conductor of the ring circuit, ohm
R_{2T} = total resistance of the protective conductor of the ring circuit, ohm
R_{1S} = resistance of the phase conductor of the spur, ohm
R_{2S} = resistance of the protective conductor of the spur, ohm.

Example 3.4

A single-phase ring circuit is run in 2.5 mm^2 70°C pvc-insulated and sheathed flat cable to BS 6004 (the protective conductor cross-sectional area being 1.5 mm^2). The total length of the ring circuit is 80 m. A spur taken from the ring is run in the same cable, the length of the spur being 12 m. Protection against indirect contact is provided by a device of a type listed earlier in this chapter.

If it is estimated that the spur is taken from the ring 30 m from the origin and $Z_E = 0.35$ ohm, what is the earth fault loop impedance for the spur?

Answer

$$y = \frac{30}{80} = 0.375$$

The relevant table for both the ring and spur is Table 54C and Column 4 of Table 3.1 applies. From that column and table, $(R_1 + R_2)/m = 23.4$ milliohms/m.

For the ring circuit:

$$(R_{1T} + R_{2T}) = \frac{23.4 \times 80}{1000} \text{ ohm} = 1.87 \text{ ohm}$$

For the spur:

$$(R_{1S} + R_{2S}) = \frac{23.4 \times 12}{1000} \text{ ohm} = 0.281 \text{ ohm}$$

Thus:

$$Z_s = 0.35 + (0.375 \times 0.625 \times 1.87) + 0.281 \text{ ohm} = 1.07 \text{ ohm}$$

The factor 0.625 in the above equation is, of course (1 – 0.375), namely (1 – y).

The alternative method, giving a pessimistically high result, is to calculate the earth fault loop impedance of the spur as if it was taken from the mid-point of the ring circuit. For the above example this would lead to:

$$Z_s = 0.35 + (0.25 \times 1.87) + 0.281 \text{ ohm} = 1.10 \text{ ohm}$$

Example 3.5

A three-phase circuit, bunched with other circuits, is run in single-core 85°C rubber-insulated cable having copper conductors, the live and protective conductors being of 16 mm^2 cross-sectional area. A device of a type not included in the earlier list provides protection against indirect contact.

If $l = 30$ m and $Z_E = 0.8$ ohm, what is the earth fault loop impedance?

Answer

Because the circuit is bunched with other circuits, Table 54C is relevant and the values of R_1 and R_2 are obtained from Table 3.3 with the factors from Table 3.4.

Thus:

$$Z_s = 0.8 + \frac{(1.53 \times 1.15 + 1.53 \times 1.15)}{1000} \times 30 = 0.906 \text{ ohm}$$

Unlike those of the previous examples the protective conductor does comply with Table 54G and therefore it is not necessary to check compliance with the adiabatic equation of Regulation 543–01–03 provided the overcurrent protective device is intended to give *overload* protection. See also Example 4.1.

Example 3.6

A 230 V single-phase circuit is run in flat two-core (with cpc) 70°C pvc-insulated and sheathed cable, the cross-sectional area of the live conductors being 10 mm^2 and that of the protective conductor being 4 mm^2. The circuit is protected by a 50 A BS 88 'gG' fuse.

If $l = 30$ m check that the circuit complies with both Table 41C and Table 41D of BS 7671 when $Z_E = 0.6$ ohm.

Answer

From Table 41D the maximum permitted earth fault loop impedance is found to be 1.09 ohm and, from Table 41C, the maximum permitted resistance of the circuit protective conductor is 0.19 ohm.

Check the earth fault loop impedance (Z_s).

As Table 54C applies, from Table 3.1 Column 4 the value of $(R_1 + R_2)/m$ is 7.73 milliohms/m.

Thus:

$$R_1 + R_2 = \frac{7.73 \times 30}{1000} \text{ ohm} = 0.232 \text{ ohm}$$

$$Z_s = (0.6 + 0.232) \text{ ohm} = 0.832 \text{ ohm}$$

The circuit therefore complies with Table 41D. Now check the resistance of the 4 mm^2 protective conductor.

Table 3.1 Column 4 can still be used, from which:

$$R_2/m = 5.53\,\text{milliohms/m}$$

Thus:

$$R_2 = \frac{5.53 \times 30}{1000}\,\text{ohm} = 0.17\,\text{ohm}$$

The circuit therefore also complies with Table 41C.

This example is illustrative of the usefulness of the alternative method permitted by Regulation 413–02–12 for circuits normally requiring a 0.4 s disconnection time, which allows that time to be extended to 5 s provided the impedance of the protective conductor does not exceed the relevant value given in Table 41C. Had the circuit been one which normally required 0.4 s disconnection which in turn meant that Z_s should not exceed 0.632 ohm (see Table 41B1) it obviously could not comply as Z_E above is itself 0.6 ohm.

The examples so far have followed the principle stated earlier. Where Table 54B applies, i.e. where the temperature of the circuit protective conductor is lower than that of the associated phase conductor, their resistances under fault conditions are calculated separately because their temperatures under those conditions are assumed to be different.

However, it will be seen from Table 3.1, for pvc insulation and a particular combination of cross-sectional areas of the phase and circuit protective conductors, that the difference between the Table 54B and Table 54C values is generally not more than 10%, the latter being the larger, similarly for Table 3.2.

Thus the designer has a further option of adopting the appropriate Table 54C value irrespective of the method of installation of the cable(s) concerned and without incurring an over-pessimistic calculated value of earth fault loop impedance. This also has the advantage of taking account of the greater temperature rise of a protective conductor which has a smaller cross-sectional area than the phase conductor.

THE MORE ACCURATE APPROACH TAKING ACCOUNT OF CONDUCTOR TEMPERATURE

Bearing in mind the fact that the formulae for determining the earth fault loop impedance are inherently approximate and that in many cases the designer has to assume the value of Z_E there is no real justification for then attempting to make a more accurate calculation of earth fault loop impedance.

However, if a more accurate calculation is felt to be necessary, one possible method is based on using the actual ambient temperature (t_a°C) and not 30°C (t_r°C) when Table 54B applies, and using the calculated conductor operating temperature (t_1°C) instead of the maximum normal operating temperature (t_p°C) in calculating the earth fault loop impedance.

When Table 54C applies, t_1°C would be used in the calculation for both the live and protective conductors.

Tables 3.3 and 3.4 have been prepared, the former giving values of resistance per metre at 20°C for conductors up to and including 35 mm^2 while the latter gives the multipliers to be applied to those values in order to determine the $(R_1 + R_2)$/m values for use in the earth fault loop impedance calculations. Table 3.5 gives factors, calculated from the first part of Table 3.4, for various values of t_1.

Table 3.3 Conductor resistances at 20°C in milliohms/metre.

Cross-sectional area mm^2	Copper	Aluminium
1	18.1	—
1.5	12.1	—
2.5	7.41	—
4	4.61	—
6	3.08	—
10	1.83	—
16	1.15	1.91
25	0.727	1.2
35	0.524	0.868

Table 3.4 Multipliers to be applied to resistance values of Table 3.3 to give R_1/m and $(R_1 + R_2)$/m in milliohms/metre for impedance calculations.

		70°C pvc insulation	85°C rubber insulation	XLPE insulation
Protective device listed in Appendix 3 of BS 7671				
Where Table 54B applies	Protective conductor	$0.92 + 0.004t_a$	$0.92 + 0.004t_a$	$0.92 + 0.004t_a$
	Phase conductor	$0.92 + 0.004t_1$	$0.92 + 0.004t_1$	$0.92 + 0.004t_1$
Where Table 54C applies	Both conductors	$0.92 + 0.004t_1$	$0.92 + 0.004t_1$	$0.92 + 0.004t_1$
Protective device *not* listed in Appendix 3 of BS 7671				
Where Table 54B applies	Protective conductor	$1.24 + 0.002t_a$ (1.30)	$1.36 + 0.002t_a$ (1.42)	$1.42 + 0.002t_a$ (1.48)
	Phase conductor	$1.24 + 0.002t_1$ (1.38)	$1.36 + 0.002t_1$ (1.53)	$1.42 + 0.002t_1$ (1.60)
Where Table 54C applies	Both conductors	$1.24 + 0.002t_1$ (1.38)	$1.36 + 0.002t_1$ (1.53)	$1.42 + 0.002t_1$ (1.60)

The values given in brackets in the lower half of the table are based on an ambient temperature of 30°C and an initial temperature which is equal to the maximum permitted normal operating temperature.

Example 3.7

A circuit is run in single-core 85°C rubber-insulated cables having copper conductors clipped direct, not grouped with the cables of other circuits.

Protection against indirect contact is provided by a device of a type listed earlier in this chapter, other than a semi-enclosed fuse. The phase conductor has a cross-sectional area of 10 mm^2 while the protective conductor has a cross-sectional area of 4 mm^2.

If $t_a = 45°C$, $t_l = 70°C$, $l = 40$ m and $Z_E = 0.35$ ohm, what is the earth fault loop impedance?

Answer

For the protective conductor, because Table 54B applies:

$$R_2 = \frac{4.61 \times [0.92 + (0.004 \times 45)] \times 40}{1000} \text{ ohm} = 0.203 \text{ ohm}$$

For the phase conductor:

$$R_1 = \frac{1.83 \times [0.92 + (0.004 \times 70)] \times 40}{1000} \text{ ohm} = 0.088 \text{ ohm}$$

Thus:

$$Z_s = (0.35 + 0.088 + 0.203) \text{ ohm} = 0.641 \text{ ohm}$$

Had the simpler approach been used, from Column 6 of Table 3.1 $(R_1 + R_2)/m = 7.10$ milliohms/m.

$$Z_s = 0.35 + \left(\frac{7.10 \times 40}{1000}\right) \text{ ohm} = 0.634 \text{ ohm}$$

Note that in this particular case the simpler approach gives a marginally lower calculated Z_s, because the higher temperature of the protective conductor has more effect than the low temperature of the phase conductor.

Example 3.8

A single-phase circuit is protected against overload by a BS 3036 semi-enclosed fuse. The circuit is run in flat twin-core 70°C pvc-insulated and sheathed cable (with cpc) having copper conductors. The cross-sectional area of the phase conductor is 6 mm^2 and that of the protective conductor is 2.5 mm^2.

If $t_a = t_r = 30°C$, $l = 50$ m and $Z_E = 0.35$ ohm, what is the earth fault loop impedance of the circuit?

Answer

As indicated in Chapter 2, because the BS 3036 fuse is providing overload protection, t_l cannot exceed 51°C when $t_a = 30°C$.

Table 54C applies and from Table 3.4 the multiplying factor is $0.92 + (0.004 \times 51) = 1.124$.

From Table 3.3, $(R_1 + R_2)/m$ at 20°C = (3.08 + 7.41) milliohms/m.

Thus:

$$Z_s = 0.35 + \frac{1.124 \times (3.08 + 7.41) \times 50}{1000} \text{ ohm} = 0.940 \text{ ohm}$$

Had the simpler method been used Z_s would have been 0.98 ohm. This demonstrates that the more accurate method for calculating Z_s gives only a slight difference in $(R_1 + R_2)$ and hence in Z_s.

Example 3.9

A three-phase circuit is run in single-core 70°C pvc-insulated and sheathed cables clipped direct and not grouped with cables of other circuits. The cross-sectional area of the phase conductors is 25 mm^2 and that of the protective conductor is 10 mm^2. A device of a type not included in the earlier list provides protection against indirect contact.

If $t_a = 5°C$, $t_l = 70°C = t_p$ (i.e. the maximum permitted normal operating temperature), $Z_E = 0.8$ ohm and $l = 20$ m, what is the earth fault loop impedance?

Answer

Table 54B applies so that from Table 3.4 the multiplier for the protective conductor is:

$$1.24 + (0.002 \times 5) = 1.25$$

and the multiplier for the phase conductor is:

$$1.24 + (0.002 \times 70) = 1.38.$$

Using the appropriate resistance per metre values from Table 3.3:

$$R_1 = \frac{1.38 \times 0.727 \times 20}{1000} \text{ ohm} = 0.0201 \text{ ohm}$$

$$R_2 = \frac{1.25 \times 1.83 \times 20}{1000} \text{ ohm} = 0.0458 \text{ ohm}$$

Thus:

$$Z_s = (0.8 + 0.0201 + 0.0458) \text{ ohm} = 0.866 \text{ ohm}$$

The only advantage gained by using this method in this particular case is a very marginal reduction in the calculated value of R_2. Had the simple method been used which assumes $t_a = t_r = 30°C$ R_2 would have been:

$$\frac{1.30 \times 1.83 \times 20}{1000} \text{ ohm} = 0.0476 \text{ ohm}$$

Table 3.5 Multipliers corresponding to $t_a = 30°C$ and various values of t_1 to be applied to the resistance values of Table 3.3.

Temperature t_1	70°C pvc insulation	85°C rubber insulation	XLPE insulation
40	1.08	1.08	1.08
50	1.12	1.12	1.12
60	1.16	1.16	1.16
70	—	1.20	1.20
80	—	—	1.24

To complete this particular section it is of interest to show further how the tabulated mV/A/m values (or the $(mV/A/m)_r$ values for larger cables) given in Appendix 4 of BS 7671 can be used to determine conductor resistances at any temperature. Some readers might prefer this approach to that using the tables given earlier in this chapter.

For single-phase circuits the resistance of *one conductor* in ohms at any temperature t_x°C is given by:

$$\frac{\text{Tabulated mV/A/m}}{2 \times 1000} \times \left(\frac{230 + t_x}{230 + t_p}\right) \times \text{length of circuit in metres}$$

For three-phase circuits the resistance of *one conductor* in ohms at any temperature t_x°C is given by:

$$\frac{\text{Tabulated mV/A/m}}{\sqrt{3} \times 1000} \times \left(\frac{230 + t_x}{230 + t_p}\right) \times \text{length of circuit in metres}$$

Example 3.10

A single-phase circuit is run in 16 mm^2 two-core (with cpc) flat 70°C pvc-insulated and sheathed cable to BS 6004, the cross-sectional area of the protective conductor being 6 mm^2. Protection against indirect contact is provided by a device of a type listed earlier in this chapter, other than a semi-enclosed fuse.

If $l = 40$ m and $t_x = 55°C$ calculate $(R_1 + R_2)$ at that temperature.

Answer

Table 4D2B is the relevant table and $t_p = 70°C$.

From Column 3 of that table tabulated mV/A/m for 16 mm^2 is 2.8 milliohms/m and tabulated mV/A/m for 6 mm^2 is 7.3 milliohms/m.

$$R_1 = \frac{2.8}{2 \times 1000} \times \left(\frac{230 + 55}{230 + 70}\right) \times 40 \text{ ohm} = 0.0532 \text{ ohm}$$

$$R_2 = \frac{7.3}{2 \times 1000} \times \left(\frac{230 + 55}{230 + 70}\right) \times 40 \text{ ohm} = 0.1387 \text{ ohm}$$

Therefore $(R_1 + R_2) = 0.192$ ohm

Example 3.11

A single-phase circuit is run in single-core non-armoured 85°C rubber-insulated cables not bunched with cables of other circuits. The cross-sectional area of the live conductors is 16 mm^2 and that of the circuit protective conductor is 10 mm^2. The conductors are of copper. The ambient temperature is 50°C and it has been calculated that the live conductors under normal load operate at 72°C. A device of a type not included in the earlier list provides protection against indirect contact.

If $l = 45$ m calculate $(R_1 + R_2)$ under earth fault conditions for the purpose of determining the earth fault loop impedance of the circuit.

Answer

Table 4F1B is the relevant table from which the mV/A/m for 16 mm^2 is 2.9 milliohms/m and for 10 mm^2 is 4.6 milliohms/m.

From Table 54B it is seen that for 85°C rubber insulation the maximum permitted final temperature is 220°C.

Thus for the phase conductor the average temperature under earth fault conditions is

$$\frac{72 + 220}{2}\,^{\circ}\text{C} = 146^{\circ}\text{C}$$

R_1 is therefore:

$$\frac{2.9}{2 \times 1000} \times \left(\frac{230 + 146}{230 + 85}\right) \times 45\,\text{ohm} = 0.078\,\text{ohm}$$

For the protective conductor the average temperature under earth fault conditions is

$$\frac{50 + 220}{2}\,^{\circ}\text{C} = 135^{\circ}\text{C}$$

R_2 is therefore:

$$\frac{4.6}{2 \times 1000} \times \left(\frac{230 + 135}{230 + 85}\right) \times 45\,\text{ohm} = 0.120\,\text{ohm}$$

Therefore $(R_1 + R_2) = 0.198$ ohm

Example 3.12

A three-phase circuit is run in 16 mm^2 multi-core non-armoured 70°C pvc-insulated cable having aluminium conductors. Protection against indirect contact is provided by a device of a type listed earlier in this chapter.

If $l = 50$ m, $t_a = 15$°C and the calculated conductor temperature under normal load (t_1°C) is 45°C calculate $(R_1 + R_2)$ under earth fault conditions for the purpose of determining the earth fault loop impedance of the circuit.

Answer

Table 4K2B is the relevant table and from Column 4 of the volt drop section mV/A/m is found to be 3.9 milliohms/m.

The assumed temperature under earth fault conditions for a phase conductor and the circuit protective conductor is 45°C.

Because Table 54C applies, the ambient temperature is not required for the calculations. $(R_1 + R_2)$ is therefore:

$$\frac{2 \times 3.9}{\sqrt{3} \times 1000} \times \left(\frac{230 + 45}{230 + 70}\right) \times 50\,\text{ohm} = 0.206\,\text{ohm}$$

In the previous examples a value of Z_E has been given without its resistance and reactance components. Where these components are known the earth fault loop impedance is calculated as follows.

Where the circuit concerned has conductors of cross-sectional area not exceeding $35\,\text{mm}^2$:

$$Z_s = \sqrt{(R_E + R_1 + R_2)^2 + X_E^2}\ \text{ohm}$$

or where the circuit concerned has conductors of cross-sectional area exceeding $35\,\text{mm}^2$:

$$Z_s = \sqrt{(R_E + R_1 + R_2)^2 + (X_E + X_1 + X_2)^2}\ \text{ohm}$$

In both the above expressions:

R_E = resistance component of Z_E, ohm
X_E = reactance component of Z_E, ohm

Where the supply is three-phase, R_E and X_E are the per phase values.

CALCULATIONS TAKING ACCOUNT OF TRANSFORMER IMPEDANCE

Example 3.13

A single-phase circuit is run in multi-core non-armoured 70°C pvc-insulated cable having copper conductors of $95\,\text{mm}^2$ cross-sectional area, its route length $l = 20\,\text{m}$. This circuit is fed from a 50 kVA distribution transformer having an internal resistance of 0.027 ohm and an internal reactance of 0.050 ohm (values obtained from the transformer manufacturer).

What is the earth fault loop impedance of the circuit at the normal operating temperature?

Answer

In order to determine the earth fault loop impedance it is necessary to obtain the resistance and reactance values for the cable from the manufacturer, adjusting the former to the normal operating temperature.

As no information is given regarding the actual operating temperature of the phase conductor, the calculations have to be based on assuming that the temperature is the maximum permitted normal operating temperature (70°C). The same temperature is assumed for the protective conductor.

One can use the mV/A/m values in the volt drop section of the tables in Appendix 4 of BS 7671 as follows:

$$(R_1 + R_2) = \frac{(\text{mV/A/m})_r \times 1}{1000} \text{ ohm}$$

$$(X_1 + X_2) = \frac{(\text{mV/A/m})_x \times 1}{1000} \text{ ohm}$$

For the present case Table 4D2B applies and from this the $(\text{mV/A/m})_r$ is found to be 0.47 milliohms/m and the $(\text{mV/A/m})_x$ is 0.155 milliohms/m.

Thus:

$$(R_1 + R_2) = \frac{0.47 \times 20}{1000} \text{ ohm} = 0.0094 \text{ ohm}$$

$$(X_1 + X_2) = \frac{0.155 \times 20}{1000} \text{ ohm} = 0.0031 \text{ ohm}$$

$$Z_s = \sqrt{(0.027 + 0.0094)^2 + (0.050 + 0.0031)^2} \text{ ohm}$$

$$= \sqrt{0.00132 + 0.0028} \text{ ohm}$$

$$= 0.064 \text{ ohm}$$

Example 3.14

A three-phase circuit is run in multicore non-armoured 70°C pvc-insulated cable having aluminium conductors of 70 mm^2, its route length $l = 45$ m. This circuit is fed from a three-phase transformer having an internal resistance of 0.005 ohm per phase and an internal reactance of 0.017 ohm per phase (values obtained from the transformer manufacturer).

What is the earth fault loop impedance at the remote end of the circuit?

Answer

As with the previous example, it is necessary to obtain the resistance and reactance values for the cable from the manufacturer, or alternatively one can use the mV/A/m values in the volt drop section of the tables in Appendix 4 of BS 7671.

Again as no information is given regarding the actual operating temperature of the conductors the calculations have to be based on assuming that the temperature is the maximum permitted normal operating temperature (70°C).

Because the circuit is three-phase:

$$(R_1 + R_2) = \frac{2 \times (\text{mV/A/m})_r \times 1}{\sqrt{3} \times 1000} \text{ ohm}$$

$$(X_1 + X_2) = \frac{2 \times (\text{mV/A/m})_x \times 1}{\sqrt{3} \times 1000} \text{ ohm}$$

For the present case Table 4K2B applies and from this the $(\text{mV/A/m})_r$ is found to be 0.9 milliohms/m and the $(\text{mV/A/m})_x$ is 0.140 milliohms/m. Thus:

$$(R_1 + R_2) = \frac{2 \times 0.9 \times 45}{\sqrt{3} \times 1000} \text{ ohm} = 0.0468 \text{ ohm}$$

$$(X_1 + X_2) = \frac{2 \times 0.14 \times 45}{\sqrt{3} \times 1000} \text{ ohm} = 0.0073 \text{ ohm}$$

$$Z_s = \sqrt{(0.0468 + 0.005)^2 + (0.0073 + 0.017)^2} \text{ ohm}$$

$$= \sqrt{0.0027 + 0.0006} \text{ ohm}$$

$$= 0.057 \text{ ohm}$$

CALCULATIONS CONCERNING CIRCUITS FED FROM SUB-DISTRIBUTION BOARDS

Consider now the case where the final circuit concerned is fed from a sub-distribution board but again the installation is fed from the public low voltage network.

Example 3.15

In an installation which is part of a TN–C–S system, a final circuit is fed from a sub-distribution board. The distribution circuit from the main board to the sub-distribution board is run in single-core 70°C pvc-insulated cables having copper conductors, the live conductors being of 25 mm^2 cross-sectional area and the protective conductor 16 mm^2, these being in trunking with cables of other circuits. The distance between the distribution boards is 27 m. The final circuit is run in 6 mm^2 flat twin (with cpc) 70°C pvc-insulated and sheathed cable, 1 being 22 m. Protection against indirect contact is provided by a device of a type listed earlier in this chapter.

If $Z_E = 0.35$ ohm what is the earth fault loop impedance (a) at the sub-distribution board and (b) for the final circuit?

Answer

Because of the absence of information on the ambient temperature and the design currents of the circuits, the simple method has to be used.

For the distribution circuit Table 54C applies. From Table 3.1 Column 4 $(R_1 + R_2)$/m is found to be 2.25 milliohms/m.

Thus Z_s at the sub-distribution board is:

$$0.35 + \frac{(2.25 \times 27)}{1000} \text{ ohm} = 0.41 \text{ ohm}$$

For the final circuit Table 54C again applies. From Table 3.1 Column 4 $(R_1 + R_2)$ milliohms/m is found to be 12.6 milliohms/m (because the cpc is of 2.5 mm^2 cross-sectional area).

For the final circuit:

$$(R_1 + R_2) = \frac{22 \times 12.6}{1000} \text{ ohm} = 0.28 \text{ ohm}$$

The designer has two options as regards calculating the earth fault loop impedance of the final circuit. The first of these is to add to its $(R_1 + R_2)$ value the previously calculated earth fault loop impedance at the distribution board, giving

$$(0.41 + 0.28) \text{ ohm} = 0.69 \text{ ohm}$$

This is a pessimistically high value because the 0.41 ohm at the sub-distribution board has assumed that the phase and protective conductors of the distribution circuit are at the maximum normal operating temperature (70°C for pvc insulation).

The temperature of the distribution circuit cable may be less than 70°C. The second option therefore is to estimate the initial temperature of the distribution circuit, say 50°C.

Using Table 3.3:

$$(R_1 + R_2) = \frac{(0.727 + 1.15) \times (230 + 50) \times 27}{1000 \times (230 + 20)} \text{ ohm}$$

$$= 0.057 \text{ ohm}$$

Thus Z_s for the final circuit is:

$$(0.35 + 0.057 + 0.28) \text{ ohm} = 0.69 \text{ ohm}$$

In this case there is no advantage over using the simple method.

If by using the simple method the resulting calculated earth fault loop impedances do not exceed the maximum values specified in BS 7671 for the overcurrent protective devices concerned there is obviously no point whatsoever in proceeding to the more accurate approach.

It should also be remembered that the $(R_1 + R_2)$ value of the distribution circuit will be less, and in some cases considerably less, than that for a final circuit and therefore contributes little to the earth fault loop impedance of the latter.

In the following example sufficient information is available for either the simple method or the more rigorous one to be used.

Example 3.16

In an installation which is part of a TN–C–S system, a single-phase distribution circuit from the main board to a sub-distribution board is run in single-core 70°C pvc-insulated and sheathed cables, the live and protective copper conductors having a cross-sectional area of $35\,mm^2$. The cables are clipped direct and not grouped with the cables of other circuits. $I_b = 105\,A$ and $l = 30\,m$. A final circuit is taken from the sub-distribution board and is run in $4\,mm^2$ two-core (with cpc) 70°C pvc-insulated and sheathed cable cpc being of $1.5\,mm^2$ cross-sectional area. The cable is clipped direct and is not bunched with other cables. $I_b = 26\,A$ and $l = 20\,m$. Protection against indirect contact is provided by a device of a type listed earlier in this chapter.

If $Z_E = 0.35$ ohm and it can be assumed that the ambient temperature throughout the installation is the same as the reference value (30°C), determine the earth fault loop impedance (a) at the sub-distribution board and (b) for the final circuit.

Answer

Using the simple method first calculate $(R_1 + R_2)$ for the distribution circuit. Table 54B applies.

From Table 3.1 Column 3, $(R_1 + R_2)$/m is found to be 1.17 milliohms/m.

Thus:

$$R_1 + R_2 = \frac{1.17 \times 30}{1000}\ \text{ohm} = 0.035\ \text{ohm}$$

Z_s at the sub-distribution board is given by:

$$(0.35 + 0.035)\ \text{ohm} = 0.385\ \text{ohm}$$

Now calculate $(R_1 + R_2)$ for the final circuit. Table 54C applies.

From Table 3.1 Column 4, $(R_1 + R_2)$/m is found to be 20.1 milliohms/m.

Thus:

$$(R_1 + R_2) = \frac{20.1 \times 20}{1000}\ \text{ohm} = 0.402\ \text{ohm}$$

Z_s for the final circuit is:

$$(0.385 + 0.402)\,\text{ohm} = 0.787\,\text{ohm}$$

Now consider the more rigorous approach again taking the distribution circuit first.

From Table 4D1A Column 6 it is found that $I_{ta} = 141\,\text{A}$ for $35\,\text{mm}^2$ conductor cross-sectional area.

With $I_b = 105\,\text{A}$:

$$t_1 = 30 + \left(\frac{105^2}{141^2} \times 40\right)^\circ\text{C} = 52.2^\circ\text{C}$$

From Tables 3.3 and 3.4, for under fault conditions:

$$R_1 = \frac{0.524 \times [0.92 + (0.004 \times 52.2)]}{1000} \times 30\,\text{ohm} = 0.0177\,\text{ohm}$$

$$R_2 = \frac{0.524 \times [0.92 + (0.004 \times 30)]}{1000} \times 30\,\text{ohm} = 0.0163\,\text{ohm}$$

Z_s at the sub-distribution board is given by:

$$(0.35 + 0.0177 + 0.0163)\,\text{ohm} = 0.383\,\text{ohm}$$

It will be seen immediately that this value is the same as that calculated by the simple method.

As regards the final circuit itself:

From Column 6 of Table 4D2A it is found that I_{ta} for $4\,\text{mm}^2$ cross-sectional area is 36 A. Thus with $I_b = 26\,\text{A}$ the conductor operating temperature is given by:

$$30 + \left(\frac{26^2}{36^2} \times 40\right)^\circ\text{C} = 50.9^\circ\text{C}$$

Using Tables 3.3 and 3.4 then R_1 is given by:

$$4.61 \times \frac{[0.92 + (0.004 \times 50.9)]}{1000} \times 20\,\text{ohm} = 0.104\,\text{ohm}$$

and R_2 is given by:

$$12.1 \times \frac{[0.92 + (0.004 \times 50.9)]}{1000} \times 20\,\text{ohm} = 0.272\,\text{ohm}$$

Note that because the protective conductor is in the same cable as the live conductors it is assumed that the former attains the same temperature as the latter.

Z_s for the final circuit is given by:

$$(0.383 + 0.104 + 0.272)\,\text{ohm} = 0.759\,\text{ohm}$$

It is a matter of personal judgement as to whether the more rigorous approach has been worthwhile. Certainly, had the simpler approach given impedance values less than the specified maximum there would be no advantage in then using the more rigorous method, as already indicated.

A simple approach to calculating earth fault loop impedance when the protective device is one of those listed earlier is to assume that all the conductors are at their maximum permitted normal operating temperature.

Where the protective device is not one of those listed earlier the possible approaches are indicated in Figure 3.1. The following method is a compromise which has some merit where the protective device is not one of those listed earlier.

(1) To calculate the earth fault loop impedance at the sub-distribution board use the simple method based on Tables 3.1 and 3.2, and
(2) To calculate the earth fault loop impedance for the final circuit use the simple method to calculate its own $(R_1 + R_2)$ value but in calculating the contribution of the distribution circuit cable base its $(R_1 + R_2)$ on that cable being assumed to be at the maximum permitted normal operating temperature for the cable insulation concerned and that any temperature rise due to an earth fault in the final circuit can be ignored. The temperature assumed for the protective conductor of the distribution circuit will depend on whether Table 54B or Table 54C applies.

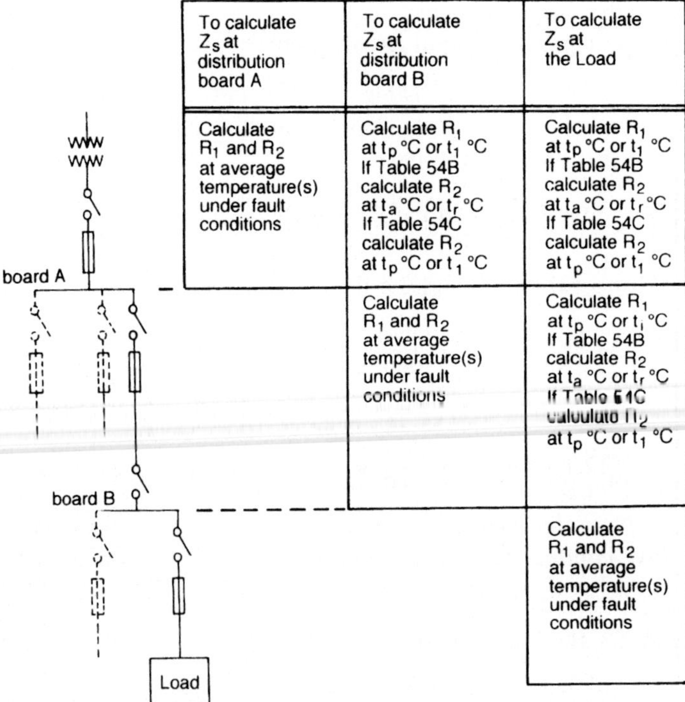

Figure 3.1 Temperatures at which conductor resistances should be calculated when determining earth fault loop impedances.

Example 3.17

In an installation which is part of a TN–S system a final circuit from a sub-distribution board is run, not bunched with other circuits, in single-core 70°C pvc-insulated and sheathed cables having copper conductors, $6\,\text{mm}^2$ cross-sectional area for the live conductors and $2.5\,\text{mm}^2$ for the protective conductor. $I_b = 33\,\text{A}$ and $l = 28\,\text{m}$.

The distribution circuit to the sub-distribution board is also run, not bunched with other circuits, in single-core 70°C pvc-insulated and sheathed cables with copper conductors of $25\,\text{mm}^2$ cross-sectional area for the live and protective conductors. For this circuit $l = 20\,\text{m}$ and $I_b = 105\,\text{A}$. Z_E is taken to be 0.8 ohm and the expected ambient temperature throughout the installation is 30°C.

What is the earth fault loop impedance for the final circuit?

Answer

Following the proposed compromise approach the $(R_1 + R_2)$ contribution of the distribution circuit to the required earth fault loop impedance is calculated assuming that the circuit live conductors are operating at 70°C and, because Table 54B applies, its protective conductor is at the ambient temperature.

For the distribution circuit, from Table 3.1 Column 3

$$(R_1 + R_2) = \frac{1.63 \times 20}{1000} = 0.0326$$

For the final circuit itself, from Table 3.1 Column 3 (because Table 54B again applies):

$$(R_1 + R_2) = \frac{11.4 \times 28}{1000}\,\text{ohm} = 0.3192\,\text{ohm}$$

Z_s for the final circuit is given by $(0.8 + 0.0326 + 0.3192)\,\text{ohm} = 1.15\,\text{ohm}$.

CALCULATIONS WHERE CONDUIT OR TRUNKING IS USED AS THE PROTECTIVE CONDUCTOR

Regulation 543–02–02 lists those items which may be used as protective conductors and amongst these are:

(i) the metallic sheaths, screens and armouring of cables
(ii) metallic conduits or other enclosures for conductors.

Item (i) includes the sheaths of mineral-insulated cables and item (ii) obviously includes trunking and ducting. All these items exhibit more than adequate thermal capacity under earth fault conditions but there is a particular problem associated with these when it comes to assessing their contribution to the earth fault loop impedance of a circuit.

For mineral-insulated cables the cross-sectional area of the sheath is greater than, and in many cases considerably greater than, that of the associated live conductor and it would be unreasonable to assume that under earth fault conditions the sheath would attain the same temperature as the live conductor.

The manufacturer of the mineral-insulated cables it is intended to use should be consulted regarding the values of $(R_1 + R_2)$ per metre under fault conditions.

Consider now the case where the circuit protective conductor is metallic conduit.

The basic problem when calculating the contribution of the conduit to earth fault loop impedance is that the impedance of the conduit varies with the magnitude of the fault currents. It rises to a peak between 50 A and 70 A and then decreases, the peak value being of the order of three times the d.c. resistance. It is therefore logical to adopt two values of the impedance of the circuit, one for fault currents up to 100 A and the other for fault currents in excess of 100 A.

A further problem with steel conduit is that its contribution to the earth fault loop impedance has both resistive and reactive components, the impedance angle being about 33°. However, since the available conduit impedances are based on empirical values and are likely to vary depending on the magnetic properties of the steel, for most purposes it is sufficiently accurate to add the conduit impedance to the conductor resistance.

The method is to first calculate the fault current using the higher of the impedance values i.e. that for fault current up to 100 A. Should the result be less than 100 A the contribution of the conduit to the earth fault loop impedance is as calculated. Should the result be greater than 100 A the conduit contribution has to be recalculated using the lower value of the impedance in the relevant table.

Tables 3.6 and 3.7 give the impedance values in milliohms/metre for light gauge and heavy gauge conduit respectively.

It is of course necessary, when determining the earth fault loop impedance, to calculate the resistance of the phase conductor and in order to do this the values given in Table 3.1 are used, if that conductor is of copper, or Table 3.2 when of aluminium. Alternatively Tables 3.3 or 3.4 may be used if 'correcting' the phase conductor resistance for operating temperature.

If the designer intends to use the method of compliance with the requirements for automatic disconnection by limiting the impedance of the protective conductor, i.e. by not exceeding the relevant value given in Table 41C of BS 7671 the values given in Table 3.8 should be used to determine what has been called the touch voltage impedance of the conduit.

The values in Tables 3.6, 3.7 and 3.8 may be assumed to be independent of temperature.

Table 3.6 Contribution of light gauge conduit to earth fault loop impedance.

Nominal conduit size mm	Conduit impedance milliohms/m	
	When $I_{ef} \leqslant 100$ A	When $I_{ef} > 100$ A
1	2	3
16	7.8	5.0
20	5.4	4.0
25	3.5	2.2
32	2.4	1.5

Table 3.7 Contribution of heavy gauge conduit to earth fault loop impedance.

Nominal conduit size mm	Conduit impedance milliohms/m	
	When $I_{ef} \leqslant 100$ A	When $I_{ef} > 100$ A
1	2	3
16	7.6	3.8
20	4.7	2.5
25	3.2	1.7
32	2.0	1.1

Table 3.8 Impedance of conduit related to Table 41C of BS 7671.

Nominal conduit size mm	Conduit impedance milliohms/m	
	Light gauge	Heavy gauge
1	2	3
16	2.8	2.1
20	2.2	1.4
25	1.5	1.1
32	1.1	0.85

If it is considered necessary to deal with the resistive and reactive components of the conduit impedance independently then values of R_2 and X_2 can be obtained from the impedances given in Tables 3.6 and 3.7 as follows:

when $I_{ef} \leqslant 100$ A $\quad R_2 = 0.83Z_2$ and $X_2 = 0.56Z_2$
when $I_{ef} > 100$ A $\quad R_2 = 0.71Z_2$ and $X_2 = 0.71Z_2$

Example 3.18

A 230 V single-phase circuit is run in 6 mm^2 single-core 70°C pvc-insulated cables, having copper conductors in 25 mm metallic light gauge conduit and protection against indirect contact is provided by the overcurrent protective device which is a 32 A BS 88 'gG' fuse.

If $l = 20$ m and $Z_E = 0.35$ ohm check that the circuit meets the maximum disconnection time of 5 s.

Answer

From Table 3.1, Column 3, $R_1 = 3.70$ milliohms/m.

From Table 3.6, Column 2, the conduit contribution to Z_s is 3.5 milliohms/m.

Thus:

$$(R_1 + Z_2) = \frac{(3.70 + 3.5)}{1000} \times 20\,\text{ohm} = 0.144\,\text{ohm}$$

$$Z_S = (0.144 + 0.35)\,\text{ohm} = 0.494\,\text{ohm}$$

$$I_{ef} = \frac{230}{0.494}\,\text{A} = 466\,\text{A}$$

As $I_{ef} > 100$ A, Z_s has to be recalculated, the conduit contribution, from Column 3 of Table 3.6, being reduced to 2.2 milliohms/m.

Thus:

$$(R_1 + Z_2) = \frac{(3.70 + 2.2)}{1000} \times 20\,\text{ohm} = 0.118\,\text{ohm}$$

$$Z_S = (0.118 + 0.35)\,\text{ohm} = 0.47\,\text{ohm}$$

The maximum value of Z_s (from Table 41D) for 5 s disconnection time is 1.92 ohm so the circuit complies.

It is important to note that if there are other circuits in the conduit, as there might well be, each circuit has to checked.

Example 3.19

A 400 V three-phase circuit is run in 4 mm^2 single-core 85°C rubber-insulated cables having copper conductors in 32 mm light gauge metallic conduit and protection against indirect contact is provided by the over-current protective devices which are 20 A Type B mcbs. If $l = 45$ m and $Z_E = 0.8$ ohm check the circuit meets the maximum disconnection time of 0.4 s.

Answer

From Table 3.1 Column 5, $R_1/\text{m} = 5.81$ milliohms/m.

From Table 3.6 Column 2, the conduit contribution to Z_s is 2.4 milli-ohms/m.

Thus:

$$(R_1 + Z_2) = \frac{(5.81 + 2.4)}{1000} \times 45 = 0.369\,\text{ohm}$$

$$Z_S = (0.369 + 0.8)\,\text{ohm} = 1.169\,\text{ohm}$$

$$\text{As } U_o = 230\,\text{V}, \quad I_{ef} = \frac{230}{1.169}\,\text{A} = 197 - \text{A}$$

Because this is greater than 100 A it is necessary to recalculate Z_s, the conduit contribution from Column 3 of Table 3.6 being reduced to 1.5 milliohms/m.

$$(R_1 + Z_2) = \frac{(5.81 + 1.5)}{1000} \times 45\,\text{ohm} = 0.329\,\text{ohm}$$

$$Z_S = (0.329 + 0.8)\,\text{ohm} = 1.129\,\text{ohm}$$

From Table 41B2 the maximum value of Z_s for a disconnection time of 0.4 s is found to be 2.4 ohm so the circuit complies.

Note that in Table 41B2 the same value of earth fault loop impedance applies to both 0.4 s and 5 s maximum disconnection times for mcbs. This is because the impedance values are based on the so-called instantaneous disconnection time of 0.1 s.

BS 7671, however, allows the designer to adopt a different value of maximum permitted earth fault loop impedance for 0.4 s disconnection compared with that for 5 s where the characteristic of the circuit breaker is such that this differentiation can be made but the examples used throughout this book are based on the 0.1 s disconnection time values.

Example 3.20

A 400 V three-phase circuit is run in 6 mm^2 single-core 70°C pvc-insulated cables having copper conductors in 16 mm light gauge metallic conduit and protection against indirect contact is provided by the overcurrent protective devices which are 25 A BS 88 'gG' fuses.

If $1 = 85$ m and $Z_E = 0.8$ ohm, check the circuit meets the maximum disconnection time of 0.4 s.

Answer

From Table 3.1 Column 3, $R_1/m = 3.70$ milliohms/m.

From Table 3.6 Column 2 the conduit contribution to Zs is 7.8 milliohms/m.

Thus:

$$(R_1 + R_2) = \frac{(3.70 + 7.8)}{1000} \times 85\,\text{ohm} = 0.98\,\text{ohm}$$

$$Z_S = (0.98 + 0.8)\,\text{ohm} = 1.78\,\text{ohm}$$

This exceeds the maximum permitted value of 1.5 ohm given in Table 41B1 for 0.4 s disconnection time.

But the earth fault current is given by:

$$I_{ef} = \frac{230}{1.78}\,\text{A} = 129\,\text{A}$$

Because this is greater than 100 A it is necessary to recalculate Z_2 and from Table 3.6 Column 3 it is found that the original value of 7.8 milliohms/m is reduced to 5 milliohms/m with fault currents in excess of 100 A.

The 'corrected' $(R_2 + Z_2)$ is then given by:

$$\frac{(3.70 + 5)}{1000} \times 85\,\text{ohm} = 0.74\,\text{ohm}$$

and the 'corrected' Z_s is $(0.74 + 0.8)$ ohm = 1.54 ohm. The 'corrected value' is nevertheless still greater than the permitted maximum value of 1.5 ohm.

However the disconnection time could be increased to 5 s because the calculated Z_s does not exceed the value of 2.4 ohm given in Table 41D provided that the impedance of the conduit using Table 3.8 does not exceed 0.43 ohm as stated in Table 41C.

From Table 3.8 Column 2 the impedance of the conduit/metre is found to be 2.8 milliohms/m. So this impedance is

$$\frac{(2.8 \times 85)}{1000}\,\text{ohm} = 0.238\,\text{ohm}$$

Therefore the so-called 'alternative method' allows the circuit to be used.

Where trunking is used as the cpc a similar approach to that used for conduit is suggested except that instead of the changeover point being based on the earth fault current it is based on the design current for the circuit. This is because it is considered that trunking is not suitable as a cpc for circuits carrying more than 100 A unless particular attention is paid to the joints between sections of trunking. As with conduit the impedance of trunking has both resistive and reactive components but for most purposes it is considered sufficiently accurate to add the trunking impedance to the conductor resistance. Table 3.9 gives the trunking contribution to the earth fault loop impedance and Table 3.10 gives the trunking impedance relative to Table 41C.

Example 3.21

A 230 V single phase circuit is run in 16 mm^2 single-core 85°C rubber-insulated cable having copper conductors in 75 mm × 50 mm trunking. Protection against indirect contact is provided by the overcurrent protective device which is a 63 A BS 88 'gG' fuse. If l = 35 m and Z_E = 0.35 ohm check the circuit meets the maximum disconnection time of 5 s.

Answer

From Table 3.1 Column 5, R_1/m is 1.45 milliohms/m.

From Table 3.9 Column 4 the contribution to the circuit Z_s is 1.82 milliohms/m.

Thus:

$$(R_1 + Z_2) = \frac{(1.45 + 1.82)}{1000} \times 35\,\text{ohm} = 0.11\,\text{ohm}$$

$$Z_S = (0.11 + 0.35)\,\text{ohm} = 0.46\,\text{ohm}$$

As this is less than the maximum permitted value of 0.86 ohm given in Table 41D, the circuit complies. If the more accurate approach of using R_2 and X_2 is adopted then Z_s is again 0.46 ohm.

Table 3.9 Contribution of trunking to earth fault loop impedance.

Nominal trunking size mm × mm	Trunking impedance milliohms/m			
	When $I_b \leqslant 100$ A			When $I_b > 100$ A
	R_2	X_2	Z_2	Z_2
1	2	3	4	5
50 × 38	2.23	2.12	3.08	2.13
50 × 50	1.85	1.76	2.55	1.77
75 × 50	1.32	1.26	1.82	1.26
75 × 75	1.03	0.98	1.42	0.98
100 × 50	1.16	1.10	1.60	1.10
100 × 75	0.94	0.88	1.29	0.88
100 × 100	0.65	0.62	0.90	0.63
150 × 50	0.80	0.76	1.10	0.76
150 × 75	0.65	0.62	0.90	0.63
150 × 100	0.48	0.46	0.66	0.46
150 × 150	0.38	0.36	0.52	0.37

Table 3.10 Impedance of trunking relating to Table 41C of BS 7671.

Nominal trunking size mm × mm	Trunking impedance milliohms/m
1	2
50 × 38	1.06
50 × 50	0.88
75 × 50	0.63
75 × 75	0.49
100 × 50	0.55
100 × 75	0.44
100 × 100	0.31
150 × 50	0.38
150 × 75	0.31
150 × 100	0.23
150 x 150	0.18

CALCULATIONS WHERE CABLE ARMOURING IS USED AS THE PROTECTIVE CONDUCTOR

Finally in this present chapter consider the case where the cable armouring is used as the cpc.

BS 6346 for pvc-insulated cables and BS 5467 for cables with XLPE insulation give resistance values based on 20°C for both the live conductors and the steel wire armouring. These resistance values for live conductors of up to and including 35 mm^2 cross-sectional area are given in Table 3.3 for both copper and aluminium.

The resistance values for the steel wire armouring are given in Tables 3.11 to 3.14 and it will be seen that for cables of 50 mm^2 and greater cross-sectional area an armour reactance of 0.3 milliohms/m has been given.

To determine the contribution of the live conductor and the armouring to the earth fault loop impedance, the conductor resistance is corrected to the assumed temperature under earth fault conditions. Similarly the armour resistance is corrected to 60°C for 70°C pvc-insulated cables and 80°C for cables with XLPE insulation and the temperatures are assumed to remain the same under earth fault conditions. The reactance values are independent of temperature.

For cables of 50 mm^2 and greater cross-sectional area it is suggested that the mV/A/m values given in Appendix 4 of BS 7671 are used, as shown in earlier examples, to determine the contribution of the phase conductor resistance and reactance to the earth fault loop impedance.

Table 3.11 Impedance of steel wire armouring for 70°C pvc-insulated cables to BS 6346 having stranded copper conductors at 20°C in milliohms/metre.

Conductor cross-sectional area mm^2	Two-core		Three-core		Four-core (equal)		Four-core (reduced neutral)	
	R	X	R	X	R	X	R	X
1	2	3	4	5	6	7	8	9
1.5	10.7	—	10.2	—	9.5	—	—	—
2.5	9.1	—	8.8	—	7.9	—	—	—
4	7.5	—	7.0	—	4.6	—	—	—
6	6.8	—	4.6	—	4.1	—	—	—
10	3.9	—	3.7	—	3.4	—	—	—
16	3.5	—	3.2	—	2.2	—	—	—
25	2.6	—	2.4	—	2.1	—	2.1	—
35	2.4	—	2.1	—	1.9	—	1.9	—
50	2.1	0.3	1.9	0.3	1.3	0.3	1.7	0.3
70	1.9	0.3	1.4	0.3	1.2	0.3	1.2	0.3
95	1.3	0.3	1.2	0.3	0.98	0.3	1.0	0.3
120	1.2	0.3	1.1	0.3	0.71	0.3	0.73	0.3
150	1.1	0.3	0.74	0.3	0.65	0.3	0.67	0.3
185	0.78	0.3	0.68	0.3	0.59	0.3	0.60	0.3
240	0.69	0.3	0.60	0.3	0.52	0.3	0.54	0.3
300	0.63	0.3	0.54	0.3	0.47	0.3	0.49	0.3
400	0.56	0.3	0.49	0.3	0.34	0.3	0.35	0.3

Table 3.12 Impedance of steel wire armouring of 70°C pvc-insulated cables to BS 6346 having sector shaped solid aluminium conductors at 20°C in milliohms/metre.

Conductor cross-sectional area mm^2	Two-core		Three-core		Four-core	
	R	X	R	X	R	X
1	2	3	4	5	6	7
16	3.7	—	3.4	—	2.4	—
25	2.9	—	2.5	—	2.3	—
35	2.7	—	2.3	—	2.0	—
50	2.4	0.3	2.0	0.3	1.4	0.3
70	2.1	0.3	1.4	0.3	1.3	0.3
95	1.5	0.3	1.3	0.3	1.1	0.3
120	—	—	1.2	0.3	0.78	0.3
150	—	—	0.82	0.3	0.71	0.3
185	—	—	0.73	0.3	0.64	0.3
240	—	—	0.65	0.3	0.57	0.3
300	—	—	0.59	0.3	0.52	0.3

When the designer intends to use the method of compliance with the requirements for automatic disconnection by limiting the impedance of the protective conductor, i.e. by not exceeding the relevant value given in Table 41C of BS 7671, the armour resistance, adjusted for temperature as above is used and its reactance ignored.

Example 3.22

A single-phase circuit is run in two-core armoured 70°C pvc-insulated cable having copper conductors of 4 mm^2 cross-sectional area and protected by a BS 88 fuse.

If $l = 34$ m and $Z_E = 0.35$ ohm, calculate the earth fault loop impedance.

Answer

From Table 3.3 Column 2, $R_1/m = 4.6$ milliohms/m.

R_1 at the average fault temperature is given by:

$$R_1 = \frac{4.61 \times 34 \times 1.2}{1000} \text{ ohm} = 0.188 \text{ ohm}$$

From Table 3.11 Column 2, the armour resistance is 7.5 milliohms/m at 20°C.

The armour resistance at 60°C is given by:

$$\frac{7.5 \times 34 \times 1.18}{1000} \text{ ohm} = 0.3 \text{ ohm}$$

The earth fault loop impedance is given by:

$$Z_s = (0.35 + 0.188 + 0.3)\,\text{ohm} = 0.838\,\text{ohm}$$

Alternatively R_1 could have been found more directly from Table 3.1 Column 4 which gives the R_1/m value of 5.53 milliohms/m. From which (at average fault temperature):

$$R_1 = \frac{5.53 \times 34}{1000}\,\text{ohm} = 0.188\,\text{ohm}$$

This is the same value as before.

Table 3.13 Impedance of steel wire armouring for cables to BS 5467 having XLPE insulation and stranded copper conductors at 20°C in milliohms/metre.

Conductor cross-sectional area mm^2	Two-core		Three-core		Four-core (equal)		Four-core (reduced neutral)	
	R	X	R	X	R	X	R	X
1	2	3	4	5	6	7	8	9
1.5	9.4	—	9.1	—	8.5	—	—	—
2.5	8.8	—	8.2	—	7.7	—	—	—
4	7.9	—	7.5	—	6.8	—	—	—
6	7.0	—	6.6	—	4.3	—	—	—
10	6.0	—	4.0	—	3.7	—	—	—
16	3.8	—	3.6	—	3.2	—	—	—
25	3.7	—	2.5	—	2.3	—	2.3	—
35	2.5	—	2.3	—	2.0	—	2.1	—
50	2.3	0.3	2.0	0.3	1.8	0.3	1.9	0.3
70	2.0	0.3	1.8	0.3	1.2	0.3	1.3	0.3
95	1.4	0.3	1.3	0.3	1.1	0.3	1.1	0.3
120	1.3	0.3	1.2	0.3	0.76	0.3	0.96	0.3
150	1.2	0.3	0.78	0.3	0.68	0.3	0.71	0.3
185	0.82	0.3	0.71	0.3	0.61	0.3	0.63	0.3
240	0.73	0.3	0.63	0.3	0.54	0.3	0.56	0.3
300	0.67	0.3	0.58	0.3	0.49	0.3	0.52* 0.49**	0.3
400	0.59	0.3	0.52	0.3	0.35	0.3	0.46	0.3

* with 150 mm^2 neutral
** with 185 mm^2 neutral

The factors for adjusting the 20°C resistance values to other temperatures for earth fault loop impedance calculations are:

Phase conductors – adjust to maximum permitted normal operating temperature.
1.20 for 70°C pvc insulated conductors
1.28 for 90°C XLPE insulated conductors
Phase conductors – adjust to average of working and maximum permitted final temperature.
1.38 for 70°C pvc insulated conductors, 1.34 for conductors greater than 300 mm^2
1.60 for 90°C XLPE insulated conductors
Steel-wire armour
1.18 for the steel-wire armour of 70°C pvc-insulated cables
1.27 for the steel-wire armour of 90°C XLPE insulated cables

Table 3.14 Impedance of steel wire armouring for cables to BS 5467 having XLPE insulation and solid aluminium conductors at 20°C in milliohms/metre.

Conductor cross-sectional area mm^2	Two-core		Three-core		Four-core	
	R	X	R	X	R	X
1	2	3	4	5	6	7
16	3.9	—	3.7	—	3.4	—
25	3.1	—	2.7	—	2.4	—
35	2.9	—	2.5	—	2.2	—
50	2.6	0.3	2.2	0.3	1.9	0.3
70	2.3	0.3	1.9	0.3	1.3	0.3
95	1.6	0.3	1.3	0.3	1.2	0.3
120	—	—	1.2	0.3	0.82	0.3
150	—	—	0.86	0.3	0.74	0.3
185	—	—	0.76	0.3	0.66	0.3
240	—	—	0.69	0.3	0.59	0.3
300	—	—	0.63	0.3	0.54	0.3

Example 3.23

A three-phase circuit is run in 16 mm^2 three-core armoured 70°C pvc-insulated cable having aluminium conductors. The protection against indirect contact is provided by the overcurrent protective device which is a 50 A Type B mcb (TP & N).

If $l = 42$ m, $R_E = 0.25$ ohm and $X_E = 0.20$ ohm check that the circuit earth fault loop impedance does not exceed 0.96 ohm as given in Table 41B2.

Answer

From Table 4K4B Column 4, mV/A/m = 3.9 milliohms/m (at 70°C).

$$R_1 = \frac{3.9 \times 42}{\sqrt{3} \times 1000} \text{ ohm} = 0.095 \text{ ohm}$$

From Table 3.12 Column 4:

$$R_2 \text{ at } 60°C = \frac{3.4 \times 1.18 \times 42}{1000} \text{ ohm} = 0.169 \text{ ohm}$$

Thus:

$$Z_s = \sqrt{(0.25 + 0.095 + 0.169)^2 + 0.20^2} \text{ ohm}$$

$$= 0.551 \text{ ohm}$$

The circuit therefore complies.

Example 3.24

A single-phase circuit is run in 50 mm^2 two-core armoured 70°C pvc-insulated cable having copper conductors. The protection against indirect contact is provided by the overcurrent protective device which is a 160 A BS 88 'gG' fuse.

If l = 37 m, $R_E = 0.07$ ohm and $X_E = 0.10$ ohm check that the circuit earth fault loop impedance does not exceed 0.27 ohm (as given in Table 41D for 5 s disconnection time).

Answer

From Table 4D4B Column 3, $(\text{mV/A/m})_r = 0.93$ milliohms/m at 70°C and $(\text{mV/A/m})_x = 0.165$ milliohms/m.

$$R_l = \frac{0.93 \times 37}{2 \times 1000} \text{ ohm} = 0.017 \text{ ohm}$$

$$X_l = \frac{0.165 \times 37}{2 \times 1000} \text{ ohm} = 0.003 \text{ ohm}$$

From Table 3.11 Columns 2 and 3 for the armouring:

$$R_2/\text{m} = 2.1 \text{ milliohms/m at } 20°\text{C}$$

$$R_2 \text{ at } 60°\text{C} = \frac{2.1 \times 1.18 \times 37}{1000} \text{ ohm} = 0.09 \text{ ohm}$$

$$X_2/\text{m} = 0.3 \text{ milliohms/m}$$

$$X_l = \frac{0.30 \times 37}{1000} \text{ ohm} = 0.011 \text{ ohm}$$

Thus:

$$Z_s = \sqrt{(0.07 + 0.017 + 0.09)^2 + (0.10 + 0.003 + 0.011)^2} \text{ ohm}$$

$$= 0.211 \text{ ohm}$$

The circuit therefore complies.

Chapter 4

Calculations Concerning Protective Conductor Cross-sectional Areas

The designer having calculated the earth fault loop impedances for the various circuits in an installation must then check that the calculated values do not exceed the maximum specified in BS 7671.

In those cases where the protective device concerned is not one for which BS 7671 gives a maximum value, the calculated value must be such that the specified maximum disconnection time for the circuit is not exceeded.

The next stage is to check that the protective conductors meet the thermal requirements prescribed in Chapter 54 of BS 7671. The key regulation is Regulation 543–01–01 which states that every protective conductor, other than any equipotential bonding conductor, shall either be subject to calculation using the adiabatic equation of Regulation 543–01–03 or selected in accordance with Table 54G.

However, if the overcurrent protective device, in addition to providing protection against indirect contact is providing only short circuit protection but not overload protection, the adiabatic equation must be used if the earth fault current is expected to be less than the short circuit current. This requirement applies even if the protective conductor cross-sectional area does comply with Table 54G.

Example 4.1 illustrates this particular point.

Example 4.1

A single-phase 230 V circuit is 30 m long and is run in a 4 mm^2 circular 70°C pvc-insulated and sheathed cable clipped direct having copper conductors. The circuit is taken from a sub-distribution board where the calculated prospective short circuit current is 8000 A. The overcurrent protective device is a 63 A BS 88 ‘gG’ fuse intended to provide short circuit protection only and protection against indirect contact, the load being an item of fixed equipment. The earth fault loop impedance at the sub-distribution board has been calculated to be 0.45 ohm.

Check that the circuit complies with Regulation 543–01–01.

Answer

Although from Table 4D2A Column 6, I_{ta} = 36 A, Regulation 434–01–01, because the fuse is intended to give short circuit protection and not overload protection, allows its nominal current to be greater than I_{ta}, which because no correction factors apply, is also I_z, the effective current-carrying capacity.

The prospective short circuit current at the sub-distribution board has been calculated to be 8000 A.

Thus:

$$Z \text{ phase/neutral} = \frac{230}{8000} \text{ ohm} = 0.029 \text{ ohm}$$

The phase/neutral impedance of the circuit itself (as will be explained in Chapter 5), can be found using Table 3.1, and as short circuit conditions are being considered Column 4 of that table applies.

The phase/neutral impedance of the circuit, therefore, is given by:

$$\frac{11.06}{1000} \times 30 \text{ ohm} = 0.332 \text{ ohm}$$

Thus the short circuit current with the fault occurring at the remote end of the circuit is given by:

$$\frac{230}{0.332 + 0.029} \text{ A} = 637 \text{ A}$$

From the time/current characteristic for the fuse, as given in Appendix 3 of BS 7671, it is found that the disconnection time is approximately 0.17 s.

From the adiabatic equation the maximum disconnection time that can be tolerated is given by:

$$t = \frac{k^2S^2}{I^2} \text{ s} = \frac{115^2 \times 4^2}{637^2} \text{ s} = 0.52 \text{ s}$$

The circuit conductors are therefore adequately protected thermally under short circuit conditions.

Consider now the earth fault condition. The $(R_1 + R_2)$ for the circuit is again 0.332 ohm because Table 54C applies. Thus the earth fault loop impedance for the circuit is (0.332 + 0.45) ohm = 0.782 ohm.

The current-using equipment has been stated to be a fixed item and assuming that the maximum disconnection time under earth fault conditions is 5 s, from Table 41D the maximum earth fault loop impedance for a 63 A BS 88 'gG' fuse is found to be 0.86 ohm. Thus as far as disconnection time is concerned the circuit complies.

The earth fault current is given by:

$$\frac{230}{0.782} \text{ A} = 294 \text{ A}$$

Again from the time/current characteristic for the 63 A fuse, the disconnection time is found to be approximately 4 s. But from the adiabatic equation the maximum disconnection time that can be tolerated is given by:

$$t = \frac{k^2S^2}{I_{ef}^2} \text{ s} = \frac{115^2 \times 4^2}{294^2} \text{ s} = 2.45 \text{ s}$$

Thus the circuit is not protected thermally in the event of an earth fault. So that although the circuit meets Table 54G, the earth fault loop impedance limitation and the requirements of Regulation 434–03–03 as regards the short circuit withstand capability of the conductors, it does not meet the thermal requirements associated with the earth fault condition. Hence the requirement in Regulation 543–01–01 that where the overcurrent protective device is providing short circuit protection and not overload protection, the designer must check compliance with the adiabatic equation.

This example is readily illustrated by Figure 4.1 which shows the adiabatic line for a 4 mm^2 conductor cross-sectional area and a k of 115 superimposed on the time/current characteristic for the fuse.

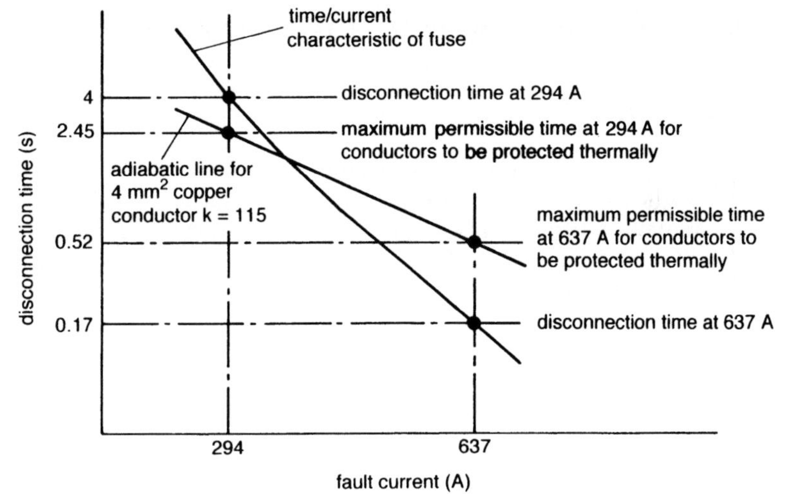

Figure 4.1 Showing how a phase conductor protected thermally under short circuit conditions is not protected when an earth fault current of lower magnitude flows.

Example 4.1 has illustrated a particular case but for the more general case where the overcurrent protective device is providing *overload* protection – with or without short circuit protection – the first stage should be to check if the protective conductor it is intended to use complies with Table 54G.

If it does so, the designer is allowed to presume that the circuit conductors are adequately protected thermally in the event of an earth fault without the need to check against the adiabatic equation. For example, with mineral-insulated cables of the multicore type the sheath cross-sectional area is considerably in excess of that demanded by Table 54G. Similarly for single-core mineral-insulated cables where the sheaths are paralleled the effective sheath cross-sectional area meets Table 54G.

BS 7671 permits the use of metallic conduit, ducting and trunking as circuit protective conductors and where these enclosures are used for this purpose it will be found that their cross-sectional areas meet Table 54G.

For instance, where the phase conductor has a cross-sectional area up to and including 16 mm^2 the minimum cross-sectional area of the enclosure has to be k_1/k_2 times that of the conductor where k_1 is the k value for the conductor from Table 54C (which is the same as that given in Table 43A) and k_2 is the value of k for the enclosure from Table 54E. Thus for 70°C pvc-insulated copper conductors in steel enclosures the cross-sectional area of the latter has to be only 2.45 times that of the conductor in order to meet Table 54G.

Similarly for armoured cables where the armour is being used as the circuit protective conductor it will he found that the cross-sectional area of the armour will comply with Table 54G in many cases.

Unfortunately, the very popular two-core and three-core 'flat' 70°C pvc-insulated and sheathed cables to BS 6004, having live conductors up to and including 16 mm^2 cross-sectional area, have protective conductors of reduced cross-sectional area which do *not* comply with Table 54G. When such cables are used the designer must *always* check the circuit design against the adiabatic equation.

CALCULATIONS WHEN THE PROTECTIVE DEVICE IS A FUSE

Example 4.2

A 230 V single-phase circuit is run in two-core 70°C pvc-insulated and sheathed cable to BS 6004, the cross-sectional area of the (copper) live conductors being 16 mm^2 and that of the protective conductor 6 mm^2.

If $l = 40$ m, $t_a = 30$°C, $Z_E = 0.35$ ohm and the circuit is protected against overload by an 80 A BS 88 'gG' fuse, check that the circuit complies with Regulation 543–01–03.

Answer

From Table 3.1 Column 4 (because Table 54C applies) it is found that $(R_1 + R_2)$/m is 5.08 milliohms/m.

$$R_1 + R_2 = \frac{5.08 \times 40}{1000}\ \text{ohm} = 0.203\ \text{ohm}$$

$$Z_s = (0.203 + 0.35)\ \text{ohm} = 0.553\ \text{ohm}$$

$$I_{ef} = \frac{230}{0.553}\ \text{A} = 416\ \text{A}$$

From the time/current characteristic the disconnection time is found to be approximately 4 s.

Check if $k^2S^2 \geqslant I_{ef}^2 t$

$$k^2S^2 = 115^2 \times 6^2\ \text{A}^2\text{s} = 476\,100\ \text{A}^2\text{s}$$

$$I_{ef}^2 t = 416^2 \times 4\ \text{A}^2\text{s} = 692\,224\ \text{A}^2\text{s}$$

Thus k^2S^2 is *not* greater than $I_{ef}^2 t$ and therefore the circuit does not comply with Regulation 543–01–03.

Example 4.3

A 400 V three-phase circuit is run in single-core non-armoured cables having copper conductors and 85°C rubber insulation, clipped direct, but bunched with cables of other circuits. The phase and neutral conductors have a cross-sectional area of $25\,mm^2$ and the protective conductor is $10\,mm^2$.

If the circuit is protected against overload by a 63 A BS 88 'gG' fuse and $l = 40\,m$, check compliance with Regulation 543–01–03 when the assumed value of Z_E is 0.25 ohm and the ambient temperature is 30°C.

Answer

From Table 3.1 Column 6 (because Table 54C applies) the $(R_1 + R_2)/m$ for the circuit is found to be 3.22 milliohms/m.

Thus:

$$R_1 + R_2 = \frac{3.22 \times 40}{1000}\ \text{ohm} = 0.129\ \text{ohm}$$

$$Z_s = (0.129 + 0.25)\ \text{ohm} = 0.379\ \text{ohm}$$

For a 400 V three-phase circuit U_o, the voltage to earth, is $400/\sqrt{3}$ V i.e. is 230 V (assuming this supply is taken from the star connected secondary windings of a transformer, the star point being earthed).

$$I_{ef} = \frac{230}{0.379}\ \text{A} = 607\ \text{A}$$

From the time/current characteristic for the fuse, the disconnection time is found to be approximately 0.18 s.

Check if $k^2S^2 \geqslant I_{ef}^2t$.

k from Table 54C is found to be 134

$$k^2S^2 = 134^2 \times 10^2\ \text{A}^2\text{s} = 1\,795\,600\ \text{A}^2\text{s}$$

$$I_{ef}^2t = 607^2 \times 0.18\ \text{A}^2\text{s} = 66\,321\ \text{A}^2\text{s}$$

Therefore the circuit does comply with Regulation 543–01–03.

Example 4.3 illustrates the particular point that when a cable is grouped with the cables of other circuits it is associated with an overcurrent protective device having a lower nominal current than that which would be used with that cable run singly. It follows that compliance with the thermal constraints under fault conditions (and the limitation in earth fault loop impedance) is less onerous when the cable is bunched.

The option is open to the designer to take account of the conductor actual operating temperature and/or the actual ambient temperature, as considered in Chapter 3 when calculating earth fault loop impedances. If this option is taken up it is then logical to adjust the k value accordingly.

The factor k is derived from the following formula:

$$k = \sqrt{\frac{Q_c(B+20)}{\rho_{20}} \ln \frac{(t_f + B)}{(t_i + B)}}$$

where: Q_c = volumetric heat capacity of conductor material (J/°C mm³)
B = reciprocal of temperature coefficient of resistivity at 0°C for the conductor (°C)
ρ_{20} = electrical resistivity of conductor material at 20°C (ohm mm)
t_i = initial temperature of conductor (°C)
t_f = final temperature of conductor (°C)
ln = $\log_e$

Values of these constants are shown in Table 4.1.

Table 4.1 Values of constants.

Material	B °C	Q_c J/°C mm³	ρ_{20} ohm mm
1	2	3	4
Copper	234.5	3.45×10^{-3}	17.241×10^{-6}
Aluminium	228	2.5×10^{-3}	28.264×10^{-6}
Lead	230	1.45×10^{-3}	214×10^{-6}
Steel	202	3.8×10^{-3}	138×10^{-6}

Figure 4.2 has been developed for copper conductors and various insulation materials, while Figure 4.3 is the equivalent for aluminium conductors. For each insulation material the k values are plotted against the initial temperature t_i°C while the final temperatures are those appropriate to the insulation material and as given in Tables 54B, 54C, 54D and 54E.

It will be seen, somewhat unexpectedly, that each graph is a straight line and the equations give, in Table 4.2, k in terms of t_i°C, the initial temperature.

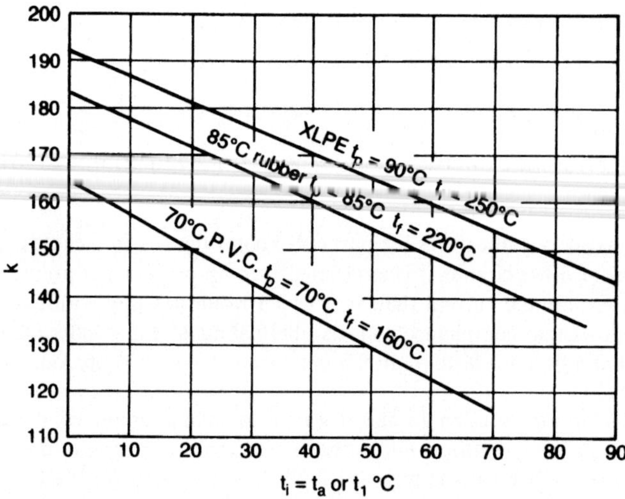

Figure 4.2 Variation of the constant k with initial temperature – copper conductors.

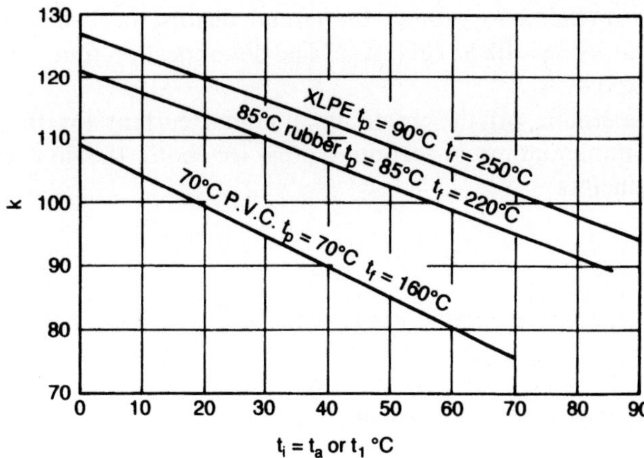

Figure 4.3 Variation of the constant k with initial temperature – aluminium conductors.

Table 4.2 Determination of k at different values of t_i°C.

Cable Insulation	For Copper Conductors	For Aluminium Conductors
1	2	3
70°C pvc	$k = 164 - 0.7t_i$	$k = 109 - 0.47t_i$
85°C rubber	$k = 183 - 0.58t_i$	$k = 121 - 0.38t_i$
XLPE	$k = 192 - 0.55t_i$	$k = 127 - 0.37t_i$

Where Table 54B applies, t_i°C $= t_r$ or t_a°C for the cpc and t_i°C $= t_p$ or t_1°C for the live conductor.
Where Table 54C applies, t_i°C $= t_p$ or t_1°C for both the cpc and live conductors.

For 70°C pvc insulation the data given in the figures and in Table 4.2 apply to cross-sectional areas not exceeding 300 mm^2.

It is not possible to indicate the potential advantage gained by using the design method which uses t_a°C instead of t_r°C, t_1°C instead of t_p°C and the corresponding k values instead of those tabulated because of the interdependence of some of the factors involved.

An example now follows which gives the results from both the simple approach and the more rigorous approach but it has to be emphasised that the comparisons made are particular to the example and should not be taken as being general.

It also shows the calculations necessary if the designer wishes to establish the minimum cross-sectional area that can be used for the circuit protective conductor.

Example 4.4

A single-phase 230 V circuit is to be run in single-core 70°C pvc-insulated and sheathed cables having copper conductors, clipped direct, and not grouped with the cables of other circuits.

$I_b = 52$ A, $t_a = 45$°C, $l = 35$ m and $Z_E = 0.70$ ohm

It is intended to protect the circuit against both overload and short circuit using a BS 88 'gG' fuse. The disconnection time for the fuse is not to exceed 5 s.

Determine (a) the minimum nominal current for the fuse, and (b) the minimum cross-sectional area for both the live and protective conductors.

Answer

First using the simple approach. As $I_n \geqslant I_b$ select from the standard nominal ratings for BS 88 'gG' fuses the next higher value greater than 52 A, i.e. 63 A.
From Table 4C1, $C_a = 0.79$.

Thus:

$$I_t = 63 \times \frac{1}{0.79}\ \text{A} = 79.7\ \text{A}$$

From Table 4D1A Column 6 it is found that the minimum conductor cross-sectional area that can be used for the live conductors is 16 mm^2 having $I_{ta} = 87$ A.

To determine the minimum protective conductor cross-sectional area may require a reiterative process. It is necessary also to check that the maximum permitted earth fault loop impedance (0.86 ohm from Table 41D) is not exceeded and then, if necessary, that the adiabatic equation is met.

As $Z_E = 0.70$ ohm, $(R_1 + R_2)$ under fault conditions $\leqslant (0.86 - 0.70)$ ohm $\leqslant 0.16$ ohm. The maximum milliohms/m value that can be tolerated is therefore:

$$\frac{0.16 \times 1000}{35}\ \text{milliohms/m} = 4.57\ \text{milliohms/m}$$

From Table 3.1 column 3 it appears that a 6 mm^2 may just be suitable for compliance with the limitation of earth fault loop impedance $(R_1 + R_2)$/m being 4.58 ohm. In this case the use of the more accurate approach taking account of conductor temperatures may be worthwhile. The following indicates the use of this alternative approach

$$t_1 = 45 + \left(\frac{52^2}{87^2} \times 40\right) = 59.3°\text{C}$$

From Table 3.3 and Table 3.4:

$$R_1 = 1.15 \times [0.92 + (0.004 \times 59.3)] = 1.33\ \text{milliohms/m}$$

and

$$R_2 = 3.08 \times [0.92 + (0.004 \times 45)] = 3.39\ \text{milliohms/m}$$

$$(R_1 + R_2)/\text{m} = 4.72\ \text{milliohms/m}$$

This demonstrates that a 6 mm^2 cpc is not acceptable. In this case the effect of the high ambient temperature more than outweighs the lower operating temperature of the phase conductor.

The minimum protective conductor to be used for compliance with the limitation of earth fault loop impedance is 10 mm^2.

With the 10 mm^2 cpc $(R_1 + R_2)/m = 3.28$ milliohms/m.

Thus:

$$R_1 + R_2 = \frac{3.28 \times 35}{1000} \text{ ohm} = 0.115 \text{ ohm}$$

$$Z_s = (0.70 + 0.115) \text{ ohm} = 0.815 \text{ ohm}$$

It remains to be checked if the 10 mm^2 protective conductor is adequately protected thermally, i.e. that it complies with the adiabatic equation.

$$I_{ef} = \frac{230}{0.815} \text{ A} = 282 \text{ A}$$

From the time/current characteristic given in Appendix 3 of BS 7671 $t = 4$ s (approximately).

Check if $k^2S^2 \geqslant I_{ef}^2t$.

k (from Table 54B) = 143.

$$k^2S^2 = 143^2 \times 10^2 \text{ A}^2\text{s} = 2\,044\,900 \text{ A}^2\text{s}$$

$$I_{ef}^2t = 282^2 \times 4 \text{ A}^2\text{s} = 318\,096 \text{ A}^2\text{s}$$

Thus k^2S^2 is greater than I_{ef}^2t so the protective conductor (and, of course, the phase conductor) is adequately thermally protected.

The next example illustrates the use of the more accurate approach where the circuit protective conductor is an integral part of the cable.

Example 4.5

A 230 V single-phase circuit is run in flat two-core cable (with cpc) 70°C pvc-insulated and sheathed, having copper conductors. The cable is clipped direct. The cross-sectional area of the live conductors is 16 mm^2 and that of the cpc is 6 mm^2. $Z_E = 0.35$ ohm and $l = 38$ m.

If the circuit is to be protected against overload and short circuit by a 63 A BS 88 'gG' fuse, $I_b = 55$ A and $t_a = 35$°C, check that the earth fault loop impedance does not exceed the specified maximum of 0.86 ohm and that the circuit complies with the requirements of Regulation 543–01–03.

Answer

The actual conductor operating temperature t_1°C is given by:

$$35 + \left(\frac{55^2}{85^2} \times 40\right) {}^\circ\text{C} = 51.7^\circ\text{C}$$

the factor 85 being the I_{ta} for the cable which is not grouped with cables of other circuits.

From Table 4.2 because Table 54C applies:

$$k = 164 - (0.7 \times 51.7) = 128.$$

From Table 3.3, $(R_1 + R_2)$/m at 20°C is (1.15 + 3.08) milliohms/m = 4.23 milliohms/m.

From Table 3.4, $(R_1 + R_2)$ is given by:

$$\frac{4.23 \times [0.92 + (0.004 \times 51.7)] \times 38}{1000} \text{ ohm} = 0.181 \text{ ohm}$$

Then $Z_s = (0.35 + 0.181)$ ohm = 0.531 ohm and therefore is less than the specified maximum.

$$I_{ef} = \frac{230}{0.531} \text{ A} = 433 \text{ A}$$

From the time current characteristic the disconnection time is found to be approximately 0.7 s.

For compliance with Regulation 543–01–03:

$$k^2S^2 = 128^2 \times 6^2 \text{ A}^2\text{s} = 589\,824 \text{ A}^2\text{s}$$

$$I_{ef}^2t = 433^2 \times 0.7 \text{ A}^2\text{s} = 131\,242 \text{ A}^2\text{s}$$

As k^2S^2 is greater than I_{ef}^2t Regulation 543–01–03 is met.

It has to be emphasised that in many cases there is no advantage in using other than the simple approach, as illustrated by Example 4.2.

Certainly for fuses up to and including those of 50 A nominal current rating it will be found that provided the maximum permitted earth fault loop impedances are not exceeded and not excessively reduced section protective conductors are used, compliance with Regulation 543–01–03 will be obtained.

CALCULATIONS WHEN THE PROTECTIVE DEVICE IS AN MCB

When protection against indirect contact is provided by an mcb (or a circuit breaker having a similar time/current characteristic) the design approach has to be different.

As indicated in Figure 4.4 the first requirement is that the earth fault loop impedance at the remote end of the circuit being protected (or at the mid point in the case of a ring circuit) has to be such that the disconnection time does not exceed 0.1 s (the specified so-called instantaneous time).

As indicated in Example 3.19, BS 7671 allows the designer to adopt a different value of maximum permitted earth fault loop impedance for 0.4 s disconnection compared with that for 5 s where the characteristic of the circuit breaker is such that

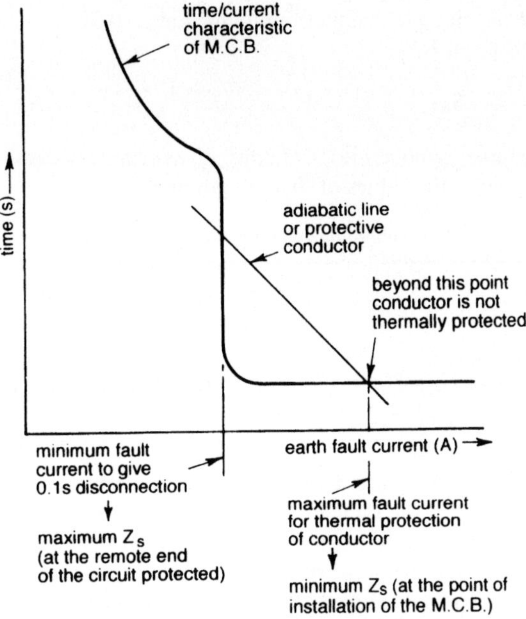

Figure 4.4 Relationship between mcb time/current characteristic, conductor adiabatic line, and minimum and maximum values for earth fault current and earth fault loop impedance.

this differentiation can be made. However, for the purposes of this analysis the assumption is made that the designer is content to use the values given in Table 41B2 for the so-called instantaneous disconnection time of 0.1 s.

As the position of the fault approaches the mcb the earth fault loop impedance decreases so the fault current increases and the disconnection time will probably be the definite minimum time which, for mcbs to BS EN 60898 is approximately 0.01 s. Thus if the earth fault loop impedance *at the point of installation of the mcb* is less than that shown in Figure 4.4 the circuit protective conductor will no longer be thermally protected, i.e. beyond the point of intersection of the time/current characteristic and the adiabatic line. The current corresponding to this point of intersection is determined by the impedances of the source and of the circuit upstream of the mcb. Thus when an mcb is used, there are both a maximum value of Z_s for the circuit and a minimum value of Z_s at the point of installation of the mcb.

The maximum value of Z_s that can be tolerated is given by

$$\frac{U_o}{aI_n} \text{ ohm}$$

where 'a' is the multiple of the nominal current (I_n) of the mcb to give 0.1 s disconnection.

For type B mcbs $a = 5$
For type C mcbs $a = 10$
For type D mcbs $a = 20$

Assuming that the definite minimum time is 0.01 s, the minimum value of Z_s *at the mcb* is given by:

$$\frac{0.10\,U_o}{kS}\text{ ohm}$$

For copper conductors of 1 mm^2 cross-sectional area and when $U_o = 230$ V Table 4.3 gives the values of this minimum Z_s.

Table 4.3 Minimum values of Z_s at the mcb.

	70°C pvc	85°C rubber	XLPE
1	2	3	4
When Table 54B applies	0.161 ohm	0.139 ohm	0.131 ohm
When Table 54C applies	0.200 ohm	0.172 ohm	0.161 ohm

As the cross-sectional area of the circuit protective conductor increases these values *decrease* in the same proportion and it is seen immediately that there should be no difficulty in meeting this requirement for installations connected to the low voltage public supply network bearing in mind the values of Z_E that occur. In other words, it is extremely unlikely that the prospective earth fault current will be greater than that corresponding to the point of intersection of the time/current characteristic and the adiabatic line.

Many present-day mcbs are of the current limiting type and, as shown in Chapter 5 when dealing with short circuit conditions, even if the calculated fault current lies beyond the point of intersection of the adiabatic line and the time/current characteristic, comparison of the k^2S^2 value for the conductor with the energy let-through (I^2t) of the mcb will, in many cases, show that the conductors *are* thermally protected.

When a circuit is protected by an mcb and that circuit is run in the standard flat two-core or three-core 70°C pvc-insulated and sheathed cables to BS 6004 there is no difficulty in complying with the thermal requirements of BS 7671 for the cpc. In fact, it will be found in many cases a much reduced cpc cross-sectional area can be used, the limiting factor in design being the need to keep within the earth fault loop maxima specified in BS 7671.

Similarly when armouring or conduit or the like is used as the cpc there should be no problem in complying with the thermal requirements under earth fault conditions.

Example 4.6

A 230 V single-phase circuit is run in 6 mm^2 two-core (with cpc) 70°C pvc-insulated and sheathed cables to BS 6004 having copper conductors (the cross-sectional area of the cpc is 2.5 mm^2). The circuit is protected by a 40 A mcb (Type B) and $l = 35$ m. It is assumed that $Z_E = 0.35$ ohm.

Check that the circuit complies with Table 41B2 and that the protective conductor satisfies the adiabatic equation of Regulation 543–01–03.

Answer

Table 54C applies, $k = 115$.

From Table 3.1 Column 4, $(R_1 + R_2)/m$ is found to be 12.6 milliohms/m.

$$R_1 + R_2 = \frac{12.6 \times 35}{1000}\text{ ohm} = 0.44\text{ ohm}$$

$$Z_s = (0.35 + 0.44)\text{ ohm} = 0.79\text{ ohm}$$

Table 41B2 indicates that the maximum permitted value of Z_s is 1.2 ohm. The circuit therefore complies with Table 41B2.

The earth fault loop impedance at the mcb must not be less than:

$$\frac{0.10 \times 230}{115 \times 2.5}\text{ ohm} = 0.08\text{ ohm}$$

This requirement is met because Z_E alone is greater. It remains to check compliance with the adiabatic equation.

The earth fault current I_{ef} is given by:

$$I_{ef} = \frac{230}{0.79}\text{ A} = 291\text{ A}$$

It is necessary to obtain the time/current characteristic from the mcb manufacturer although because the designer knows that the disconnection time is less than 0.1 s he can either use that value or from the time/current characteristics in Appendix 3 of BS 7671 it would appear the time is approximately 0.016 s.

$$k^2S^2 = 115^2 \times 2.5^2\text{ A}^2\text{s} = 82\,656\text{ A}^2\text{s}$$

$$I_{ef}^2 t = 291^2 \times 0.016\text{ A}^2\text{s} = 1355\text{ A}^2\text{s}$$

As k^2S^2 is greater than $I_{ef}^2 t$ the circuit complies with the adiabatic equation.

Example 4.7

A 400 V three-phase circuit ($U_o = 230$ V) is run in single-core non-armoured 85°C rubber-insulated cables having copper conductors, the cross-sectional area of the live conductors being 16 mm^2. The cable is installed in conduit with cables of other circuits and is protected against overload and short circuit by a TP & N 32 A Type B mcb.

If $t_a = 30$°C and $l = 48$ m determine the minimum cross-sectional area for the protective conductor for compliance with Table 41B2 and compliance with the adiabatic equation of Regulation 543–01–03. $Z_E = 0.35$ ohm.

Answer

Table 54C applies as the circuit is grouped with other circuits, $k = 134$.

From Table 41B2, $Z_s \leqslant 1.5$ ohm. Thus the maximum tolerable value of $(R_1 + R_2)$ is given by $(1.5 - 0.35)$ ohm $= 1.15$ ohm and the maximum tolerable value of $(R_1 + R_2)$ in milliohms/m is:

$$= \frac{1.15 \times 1000}{48} \text{ milliohms/m} = 23.96 \text{ milliohms/m.}$$

As the value for 16 mm^2 cross-sectional area conductors is 1.45 milliohms/m (from Table 3.1 Column 6) the maximum value for the protective conductor is (23.96 – 1.45) milliohms/m = 22.5 milliohms/m.

Inspection of Table 3.1 Column 5 shows that for compliance with Table 41B2 the minimum cross-sectional area for the protective conductor is 1.5 mm^2 having a value of 15.2 milliohms/m.

Thus:

$$(R_1 + R_2) = \frac{(1.45 + 15.2) \times 48}{1000} \text{ ohm} = 0.80 \text{ ohm}$$

$$Z_s = (0.35 + 0.80) \text{ ohm} = 1.15 \text{ ohm}$$

The circuit complies with Table 41B2.

The earth fault loop impedance at the mcb shall not be less than:

$$\frac{0.10 \times 230}{134 \times 1.5} \text{ ohm} = 0.114 \text{ ohm}$$

This requirement is met because Z_E alone is greater. It remains to check if the adiabatic equation is met.

The earth fault current I_{ef} is given by:

$$I_{ef} = \frac{230}{1.15} \text{ A} = 200 \text{ A}$$

From the time/current characteristics in Appendix 3 the disconnection time is found to be approximately 0.02 s.

$$k^2S^2 = 134^2 \times 1.5^2 \text{ A}^2\text{s} = 40\,400 \text{ A}^2\text{s}$$

$$I_{ef}^2t = 200^2 \times 0.02 \text{ A}^2\text{s} = 800 \text{ A}^2\text{s}$$

The circuit therefore meets the adiabatic equation.

Close to the mcbs the earth fault current will approach:

$$\frac{U_0}{Z_E} \text{ A} = \frac{230}{0.35} \text{ A} = 657 \text{ A}$$

The disconnection time will be 0.01 s and I_{ef}^2t becomes $657^2 \times 0.01$ A^{2}s = 4316 A^{2}s.

The circuit therefore is protected along its whole length. This last calculation, of course, is not needed as it has already been established that the earth fault loop impedance at the origin of the circuit is greater than the minimum required for a 1.5 mm^2 conductor cross-sectional area.

It should be noted that a 1.5 mm^2 cpc is also found to be sufficient if the calculations are based on the method to be used when the protective device is not one of those listed in Appendix 3 of BS 7671.

The above is an example where although the live conductors are of 16 mm^2 cross-sectional area it has been determined that the circuit protective conductor need have a cross-sectional area of only 1.5 mm^2. Because the circuit is in an enclosure of a wiring system, namely conduit, the minimum requirement of 2.5 mm^2 cross-sectional area specified in Regulation 543–01–01 does not apply.

Cases will frequently occur in practice where the cross-sectional area of the circuit protective conductor can be justifiably considerably less than that of the associated live conductors and it becomes a matter of personal preference on the part of the designer as to whether or not the cross-sectional area of the circuit protective conductor is increased.

Example 4.8

A single-phase circuit is fed from the 110 V centre-tapped earth secondary of a transformer. It is run in single-core 70°C pvc-insulated cable having copper conductors, in conduit.

If this circuit is protected by 20 A Type C mcbs, the cross-sectional area of the live conductors is 4 mm^2, l = 15 m and the internal impedance of the transformer is 0.05 ohm, determine the minimum protective conductor cross-sectional area for compliance with Table 41B2, the adiabatic equation and that the earth fault loop impedance is not below the minimum required.

Answer

From Table 41B2 the maximum Z_s is 1.20 ohm but this value applies to a circuit where $U_o = 230$ V. Here $U_o = 55$ V so maximum Z_s is:

$$\frac{55}{240} \times 1.20 \text{ ohm} = 0.275 \text{ ohm}$$

Note that the values given in Table 41B2 are based on an open circuit voltage of 240 V.

Maximum tolerable $(R_1 + R_2) = (0.275 - 0.05)$ ohm $= 0.225$ ohm

$$\text{Maximum tolerable } (R_1 + R_2)/\text{m} = \frac{0.225 \times 1000}{15} \text{ milliohms/m}$$

$$= 15.0 \text{ milliohms/m}$$

From Table 3.1 Column 4 it is found that the minimum protective conductor cross-sectional area is 2.5 mm^2.

$$(R_1 + R_2) = \frac{14.4 \times 15}{1000} \text{ ohm} = 0.22 \text{ ohm}$$

$$Z_s = (0.05 + 0.22) \text{ ohm} = 0.27 \text{ ohm}$$

The circuit meets the requirements for maximum earth fault loop impedance.

The earth fault current I_{ef} is given by:

$$I_{ef} = \frac{55}{0.27}\,A = 204\,A.$$

From the time/current characteristic the disconnection time is approximately 0.1 s. As k = 143:

$$k^2S^2 = 143^2 \times 2.5^2\,A^2s = 127\,806\,A^2s$$

$$I_{ef}^2t = 204^2 \times 0.1\,A^2s = 4162\,A^2s$$

As expected the circuit complies with the adiabatic equation. Finally check that at the origin of the circuit Z_s is not less than:

$$\frac{0.10\,U_o}{kS}\,\text{ohm}$$

i.e. is not less than

$$\frac{0.10 \times 55}{143 \times 2.5}\,\text{ohm} = 0.015\,\text{ohm}$$

As the internal impedance of the transformer is 0.05 ohm this requirement is also met.

This example shows that the previous remarks based on circuits having $U_o = 230\,V$ are equally applicable to circuits with other values of U_o. In this example the need to use a $2.5\,mm^2$ protective conductor has been dictated by the maximum value of Z_s that can be tolerated.

Some manufacturers of mcbs in their time/current characteristics do not show the definite minimum time but instead provide I^2t characteristics. Where these are available the method previously described still applies, namely that the earth fault loop impedance at the remote end of the circuit is sufficiently low to give disconnection within 0.1 s and having determined the prospective earth fault current at the origin of the circuit the total let-through energy is obtained for that current and must be less than k^2S^2.

For instance, from one manufacturer's data, even if the earth fault loop impedance at the origin of the circuit is as low as 0.12 ohm a $1\,mm^2$ protective conductor would be adequately protected thermally by an mcb of any current rating up to 50 A.

Thus, as previously stated, the minimum cross-sectional area of the protective conductor will be frequently determined by the need not to exceed the maximum specified earth fault loop impedance for 0.1 s disconnection.

All the examples so far in this chapter have concerned protection against indirect contact using overcurrent protective devices.

CALCULATIONS WHEN THE PROTECTIVE DEVICE IS AN RCCB

The remaining examples concern residual current devices.

Where the installation is part of a TN system the basic requirement is given in Regulation 413–02–16, namely that:

$$Z_sI_{\Delta n} \leqslant 50\,\text{volts}$$

where $I_{\Delta n}$ is the rated residual operating current of the rcd. For some special installations or locations, as indicated in the relevant section of Part 6 of BS 7671, this requirement is modified to:

$$Z_s I_{\Delta n} \leqslant 25 \text{ volts}$$

In both the above expressions, of course, $I_{\Delta n}$ is in amperes. Limiting consideration to residual current circuit breakers (rccbs) to BS EN 61008, the values of Z_s and $I_{\Delta n}$ encountered in practice are such that there is no difficulty in meeting the requirement given above.

The earth fault current is limited by the value of Z_s and will be of the same order as occurs when protection against indirect contact is provided by overcurrent protective devices. It should be checked that the rccb complies with Regulation 531–02–08 which requires that it is capable of withstanding without damage the thermal and mechanical stresses caused by the earth fault current.

If an rccb to BS EN 61008 does not incorporate an intentional time delay it is required by the British Standard to operate within 0.04 s at a residual current of $5I_{\Delta n}$ A and the designer merely has to check that the circuit is thermally protected for that time.

Example 4.9

A 230 V single-phase circuit is run in two-core (with cpc) 70°C pvc-insulated and sheathed cable having copper conductors of 2.5 mm^2 cross-sectional area for the live conductors and 1.5 mm^2 for the protective conductor.

If $l = 55$ m, $Z_E = 0.8$ ohm and the circuit is protected by a 30 mA rccb check compliance with the adiabatic equation of Regulation 543–01–03.

Answer

Table 54C applies. From Table 3.1:

$$(R_1 + R_2)/m = 23.4 \text{ milliohms/m}$$

$$R_1 + R_2 = \frac{23.4 \times 55}{1000} \text{ ohm} = 1.29 \text{ ohm}$$

$$Z_s = (1.29 + 0.8) \text{ ohm} = 2.09 \text{ ohm}$$

The earth fault current I_{ef} is given by:

$$I_{ef} = \frac{230}{2.09} \text{ A} = 110 \text{ A}$$

This is considerably in excess of $5I_{\Delta n}$ so the rccb will operate within 0.04 s. $k = 115$:

$$k^2S^2 = 115^2 \times 1.5^2 \text{ A}^2\text{s} = 29\,756 \text{ A}^2$$

$$I_{ef}^2 t = 110^2 \times 0.04 \text{ A}^2\text{s} = 484 \text{ A}^2\text{s}$$

The circuit therefore complies with the adiabatic equation.

This one example is sufficient to show that even had the rated residual operating current been considerably greater than 30 mA the operating time would still be less than 0.04 s. Furthermore, if one wished to, an intentional time delay could have been introduced and compliance with the adiabatic equation would still have been maintained. The maximum time would have been given by

$$\frac{29\,756}{110^2}\ \mathrm{s} = 2.46\ \mathrm{s}$$

This example suggests that if a circuit is run in single-core cables the circuit protective conductor can have a cross-sectional area considerably less than that of the associated live conductors.

Example 4.10

A 400 V ($U_o = 230$ V) three-phase circuit is run in single-core non-armoured cables having 85°C rubber insulation and copper conductors, in conduit with the cables of other circuits.

If the live conductors have a cross-sectional area of 16 mm^2, $l = 30$ m, $Z_E = 0.35$ ohm and the circuit is to be protected by a 100 mA rccb (without an intentional time delay), what is the minimum cross-sectional area that can be used for the circuit protective conductor?

Answer

Because the cables are grouped with cables of other circuits Table 54C applies and $k = 134$. Try 2.5 mm^2.

From Table 3.1 Column 6:

$$(R_1 + R_2)/m = (1.45 + 9.34)\ \text{milliohms/m} = 10.79\ \text{milliohms/m}$$

$$R_1 + R_2 = \frac{10.79 \times 30}{1000}\ \text{ohm} = 0.324\ \text{ohm}$$

$$Z_s = (0.35 + 0.324)\ \text{ohm} = 0.674\ \text{ohm}$$

The earth fault current I_{ef} is given by:

$$I_{ef} = \frac{230}{0.674}\ \mathrm{A} = 341\ \mathrm{A}$$

This is considerably in excess of $5I_{\Delta n}$ so that the disconnection time will not exceed 0.04 s.

$$k^2S^2 = 134^2 \times 2.5^2\ \mathrm{A^2s} = 112\,225\ \mathrm{A^2s}$$

$$I_{ef}^2 t = 341^2 \times 0.04\ \mathrm{A^2s} = 4651\ \mathrm{A^2s}$$

Try 1 mm^2 (which is the minimum cross-sectional area for the type of cable concerned).

$$(R_1 + R_2)/m = (1.45 + 22.8)\,\text{milliohms/m} = 24.25\,\text{milliohms/m}$$

$$R_1 + R_2 = \frac{24.25 \times 30}{1000}\,\text{ohm} = 0.73\,\text{ohm}$$

$$Z_s = (0.35 + 0.73)\,\text{ohm} = 1.08\,\text{ohm}$$

The earth fault current I_{ef} is given by:

$$I_{ef} = \frac{230}{1.08}\,A = 213\,A$$

The disconnection time again will not exceed 0.04 s.

$$k^2S^2 = 134^2 \times 1^2\,A^2s = 17\,956\,A^2s$$

$$I_{ef}^2t = 213^2 \times 0.04\,A^2s = 1815\,A^2s$$

Again, the circuit meets the adiabatic equation. It may well be that the designer would not wish to use 1 mm^2 cross-sectional area but the above shows that a cpc having it would nevertheless comply with all the relevant requirements. (See the comment at end of Example 4.11.)

For circuits where $U_o = 230$ V and the installation is part of a TN system there is no doubt that the rccb will operate within 0.04 s provided that no intentional time delay has been incorporated within the rccb as already stated. Because of this there is another way of determining immediately if the cross-sectional area of a circuit protective conductor is adequate.

From the adiabatic equation with $t = 0.04$ s:

$$k^2S^2 \geqslant I_{ef}^2 \times 0.04\,A^2s$$

But

$$I_{ef} = \frac{U_oA}{Z_s}$$

So

$$k^2S^2 \geqslant \frac{U_o^2}{Z_s^2} \times 0.04\,A^2s$$

from which:

$$Z_s \geqslant 0.2 \times \frac{U_o}{kS}\,\text{ohm}$$

Thus, for example, if $k = 115$, $U_o = 230$ V, and $S = 1\,mm^2$:

$$Z_s \geqslant \frac{0.2 \times 230}{115 \times 1}\,\text{ohm} \quad \text{i.e.} \quad Z_s \geqslant 0.40\,\text{ohm}$$

Particularly if the system is TN–S, Z_E alone is likely to be greater than this minimum value for Z_s and if the system is TN–C–S the value of Z_E may be 0.35 ohm which is a significant portion of this minimum.

If U_o is greater than 230 V again there is no problem in complying with the basic requirement and hence obtaining the 0.04 s operation time but for any given values of k and S the minimum value of Z_s will be greater than for the 230 V case.

If U_o is less than 230 V there is no problem in complying with the basic requirement and the minimum value of Z_s will be less than for the 230 V case.

Example 4.11

A 230 V single-phase circuit is run in single-core 70°C pvc-insulated cable having copper conductors, the cross-sectional area of the live conductors being 25 mm^2. It is intended to protect the circuit against indirect contact by a 0.5 A rccb.

If $l = 40$ m, $Z_E = 0.8$ ohm and there is no intentional time delay, what is the minimum cross-sectional area that can be used for the circuit protective conductor? The cable is not grouped with cables of other circuits and $t_a = 30$°C.

Answer

To comply with the adiabatic equation:

$$Z_s \geqslant \frac{0.2U_o}{kS} \text{ ohm (because t will be less than 0.04 s)}$$

Try 1 mm^2.

$$Z_s \geqslant \frac{0.2 \times 230}{143 \times 1} \text{ ohm} \qquad \text{i.e.} \qquad Z_s \geqslant 0.32 \text{ ohm}$$

As $Z_E = 0.8$ ohm then Z_s will of course exceed 0.32 ohm and the 1 mm^2 protective conductor would comply with the adiabatic equation provided the earth fault current was sufficiently great to give 0.04 s disconnection, i.e. provided $Z_s I_{\Delta n} \leqslant 50$ V.

Strictly one cannot use the values given in Table 3.1 because in Column 3 relating to those cases where Table 54B applies, as in the present example, no values are given for R_2/m.

Using Tables 3.3 and 3.4:

$$R_2/m = 18.1\,[0.92 + (0.004 \times 30)] \text{ milliohms/m} = 18.8 \text{ milliohms/m}$$

$$R_1 + R_2 = \frac{(0.872 + 18.8) \times 40}{1000} \text{ ohm} = 0.79 \text{ ohm}$$

$$Z_s = (0.8 + 0.79) \text{ ohm} = 1.59 \text{ ohm}$$

$$Z_s I_{\Delta n} = 1.59 \times 0.5 \text{ V} = 0.80 \text{ V}$$

The circuit therefore complies with the basic requirement that $Z_s I_{\Delta n} \leqslant 50$ V.

Whether the designer would use a 1 mm^2 circuit protective conductor is a matter of choice. For mechanical robustness he may decide to use a bigger conductor. In any event it must be remembered that there is a requirement in Regulation 543–01–01 that if a protective conductor

(i) is not an integral part of a cable, or
(ii) is not formed by conduit, ducting or trunking, or
(iii) is not contained in an enclosure formed by a wiring system

the cross-sectional area shall be not less than 2.5 mm^2 if sheathed or otherwise provided with mechanical protection and not less than 4 mm^2 if mechanical protection is not provided.

Now consider the case where an intentional time delay is incorporated in the rccb being used.

Example 4.12

A 400 V three-phase circuit ($U_o = 230$ V) is run in single-core 70°C pvc-insulated cables having copper conductors in trunking with the cables of other circuits. $l = 85$ m, $Z_E = 0.35$ ohm. The live conductors are of 16 mm^2 cross-sectional area. Protection is given by a 100 mA rccb incorporating a time delay such that for fault currents of 500 mA or greater the disconnection time does not exceed 4 s.

Determine the minimum cross-sectional area for the circuit protective conductor.

Answer

Try 2.5 mm^2 cross-sectional area. From Table 3.1 values (remembering that Table 54C applies):

$$R_1 + R_2 = \frac{(1.38 + 8.89) \times 85}{1000} \text{ ohm} = 0.87 \text{ ohm}$$

$$Z_s = (0.35 + 0.87) = 1.22 \text{ ohm}$$

$$Z_s I_{\Delta n} = 1.22 \times 0.1 \text{ V} = 0.122 \text{ V}$$

The circuit complies with the basic requirement that $Z_s I_{\Delta n} \leqslant 50$ V and the disconnection time will not exceed 4 s (with the incorporated time delay).

$$k = 115$$

The earth fault current I_{ef} is given by:

$$I_{ef} = \frac{230}{1.22} \text{ A} = 189 \text{ A}.$$

$$k^2S^2 = 115^2 \times 2.5^2 \text{ A}^2\text{s} = 82\,656 \text{ A}^2\text{s}$$

$$I_{ef}^2 t = 189^2 \times 4 \text{ A}^2\text{s} = 142\,884 \text{ A}^2\text{s}$$

The circuit therefore does *not* comply with the adiabatic equation but the values obtained suggest one need only increase the cross-sectional area of the circuit protective conductor to $4\,mm^2$.

Again from Table 3.1:

$$R_1 + R_2 = \frac{(1.38 + 5.53) \times 85}{1000}\ \text{ohm} = 0.59\ \text{ohm}$$

$$Z_s = (0.35 + 0.59)\ \text{ohm} = 0.94\ \text{ohm}$$

The earth fault current I_{ef} is given by:

$$I_{ef} = \frac{230}{0.94}\ A = 245\ A$$

$$k^2S^2 = 115^2 \times 4^2\ A^2s = 211\,600\ A^2s$$

$$I_{ef}^2 t = 245^2 \times 4\ A^2s = 240\,100\ A^2s$$

Again the circuit does not comply so try $6\,mm^2$.

$$R_1 + R_2 = \frac{5.08 \times 85}{1000}\ \text{ohm} = 0.43\ \text{ohm}$$

$$Z_s = (0.35 + 0.43)\ \text{ohm} = 0.78\ \text{ohm}$$

The earth fault current I_{ef} is given by:

$$I_{ef} = \frac{230}{0.78}\ A = 295\ A$$

$$k^2S^2 = 115^2 \times 6^2\ A^2s = 476\,100\ A^2s$$

$$I_{ef}^2 t = 295^2 \times 4\ A^2s = 348\,100\ A^2s$$

The minimum cross-sectional area for the circuit protective conductor is therefore $6\,mm^2$.

Because the maximum time for disconnection is known (in the present case this is 4 s) there is a method which can be adopted which immediately gives the minimum cross-sectional area for the circuit protective conductor, the assumption being made that the disconnection time *will be* 4 s.

The minimum cross-sectional area is given by:

$$\frac{\dfrac{U_o \times \sqrt{t}}{k} - (p \times 1)}{Z_E + R_1}\ mm^2$$

where $p = 0.0207$ for a maximum operating temperature of 70°C.

For the present case:

$$\text{minimum S} = \frac{\left(\dfrac{230 \times \sqrt{4}}{115}\right) - (0.0207 \times 85)}{0.35 + 0.117}\ \text{mm}^2$$

$$= 4.80\,\text{mm}^2$$

Hence, as before, the minimum conductor cross-sectional area that can be used is $6\,\text{mm}^2$. Table 4.4 gives approximate values of p for copper and aluminium conductors at 30°C, 70°C and 90°C.

Table 4.4 Approximate values of p at various temperatures.

Temperature °C	Copper	Aluminium
30	0.0179	0.0294
70	0.0207	0.0339
90	0.0221	0.0362

When the installation is part of a TT system the earth fault loop impedance will be greater than those encountered in TN systems because both the earth electrode resistance at the source of energy and that at the installation itself will be part of the earth fault loop impedance. The earth fault loop impedance can be tens or hundreds of ohms but the circuit has to comply with Regulation 413–02–20.

Compliance with this regulation again will give disconnection within 0.04 s unless the rccb has an incorporated time delay and the methods given in regard to TN systems are equally applicable to TT systems.

Chapter 5

Calculations Related to Short Circuit Conditions

The calculations related to short circuit conditions fall into two categories:

(a) those which have to be undertaken to determine whether the associated protective devices have adequate breaking capacity for compliance with Regulation 434–03–01, and
(b) those which have to be undertaken to determine whether the conductors of the circuits concerned are protected thermally for compliance with Regulation 434–03–03.

In those cases where the prospective fault currents at the origin of an installation are less than the rated breaking capacity of any of the protective devices intended to be used in the installation no further assessments of the prospective fault currents are necessary in this regard.

The calculations for item (b) are very similar to those that are undertaken in relation to earth fault conditions and it is thought to be of some benefit for the purposes of this book to deal with them first.

Before doing so it is emphasised that Regulation 435–03–02 is of considerable importance because it allows the designer to assume compliance with Regulation 434–03–03, provided that the overcurrent protective device for the circuit concerned is intended to give both overload and short circuit protection and has adequate breaking capacity.

The adiabatic equation of Regulation 434–03–03 is exactly the same as that of Regulation 543–01–03 but stated differently. This is because when considering short circuit conditions the aim usually is to check that the overcurrent protective device disconnects the circuit concerned with sufficient rapidity.

It will be noted that the k values given in Table 43A are exactly the same as those given in Table 54C. Also the assumed initial temperatures and limiting final temperatures are the same in both tables.

As with the calculation of earth fault loop impedance different assumptions are made about the conductor temperature under fault conditions depending on whether the type of protective device to be used is listed in Appendix 3 of BS 7671 or not.

If the protective device is of a type listed in Appendix 3 of BS 7671 then the conductor temperature under fault conditions can be assumed to be its normal operating temperature.

If the protective device is not of a type listed in Appendix 3 of BS 7671 then the conductor temperature under fault conditions can be assumed to be the average of its normal operating temperature and its limiting final temperature as given in Table 43A of BS 7671.

In either case the designer has two options. The first option is to assume that the normal operating temperature of the conductor is its maximum permitted normal operating temperature. The second option is to use the more rigorous method using the actual conductor temperature as the normal operating temperature, in this case the k value can be adjusted accordingly.

For consistency with the previous chapters, the nominal voltage, 230 V, has been used in the following examples. If it is known that the supply voltage is likely to be closer to the upper end of the tolerance band, then it is suggested that the open circuit voltage is used in calculating the prospective short circuit current. This will be an increase of about 4% in the prospective short circuit current.

A.C. SINGLE-PHASE CIRCUITS

The basic equation for determining the short circuit current I_{sc} is

$$I_{sc} = \frac{U_p}{\sqrt{(R_B + R_N + R_l + R_n)^2 + (X_B + X_N + X_l + X_n)^2}} \text{ A} = \frac{U_p}{Z_{pn}} \text{ A}$$

where: R_B = internal resistance of the source of energy plus the resistance of the phase conductor of the supply circuit up to the point of installation of the overcurrent protective device, ohm.

X_B = internal reactance of the source of energy plus the reactance of the phase conductor of the supply circuit up to that point, ohm.

R_N = resistance of the neutral conductor of the supply circuit up to that point, ohm.

X_N = reactance of the neutral conductor of the supply circuit up to that point, ohm.

Z_{pn} = total phase and neutral impedance.

R_l = resistance of the phase conductor of the protected circuit (as in earlier chapters), ohm

X_l = reactance of the phase conductor of the protected circuit, ohm

R_n = resistance of the neutral conductor of the protected circuit, ohm

X_n = reactance of the neutral conductor of the protected circuit, ohm

U_p = nominal voltage of the circuit, volts.

The values of R_l and R_n are at the assumed temperature under fault conditions but X_l and X_n are independent of temperature. R_B and R_N are at the normal operating temperature for the type of insulation used for the supply circuit cable(s).

Note that in some cases where there is not a neutral as such, e.g. where the circuit is supplied from a centre-tapped earthed secondary $(R_l + R_n)$ becomes $2R_l$.

In many cases, such as for installations connected directly to the low voltage public supply network R_B, X_B, R_N and X_N are not known separately and only the prospective short circuit current (I_{psc}) at the origin of the installation is known. However, this immediately gives the phase-to-neutral impedance (Z_{pn}) at the origin:

$$Z_{pn} = \frac{U_p}{I_{psc}} \text{ ohm}$$

where I_{psc} is the prospective short circuit current.

$$Z_{pn} = \sqrt{(R_B + R_N)^2 + (X_B + X_N)^2} \text{ ohm}$$

When only Z_{pn} is known the short circuit current at the remote end of the circuit concerned is given by:

$$I_{sc} = \frac{U_p}{Z_{pn} + \sqrt{(R_l + R_n)^2 + (X_l + X_n)^2}} A = \frac{U_p}{Z_{pn}} A$$

When the cross-sectional area of the conductors is less than 35 mm^2 then their reactance can be ignored and the above equation becomes:

$$I_{sc} = \frac{U_p}{Z_{pn} + R_l + R_n} A = \frac{U_p}{Z_{pn}} A$$

Having determined I_{sc} it is then necessary to obtain the corresponding disconnection time (t) from the time/current characteristic of the overcurrent protective device and to then check that this time is less than k^2S^2/I_{sc}^2 seconds.

Table 5.1 gives the values of $(R_l + R_n)/m$ for copper conductors of cross-sectional area up to and including 35 mm^2, assuming that the phase conductor and neutral conductor will have the same cross-sectional area. Table 5.2, likewise, gives the values of $(R_l + R_n)/m$ for aluminium conductors.

Table 5.1 Values of $(R_1 + R_n)/m$ for copper conductors at their normal operating temperature in milliohms/metre.

Conductor cross-sectional area mm^2	Insulation material					
	70°C pvc	60°C rubber	85°C rubber	90°C XLPE	MI 70°C sheath	MI 105°C sheath
1	2	3	4	5	6	7
1	43.4	42.0	45.6	46.3	43.7	49.1
1.5	29.0	28.1	30.5	31.0	29.2	32.8
2.5	17.8	17.2	18.7	19.0	17.9	20.1
4	11.1	10.7	11.6	11.8	11.1	12.5
6	7.39	7.15	7.76	7.88	7.44	8.35
10	4.39	4.25	4.61	4.68	4.42	4.96
16	2.76	2.67	2.90	2.94	2.78	3.12
25	1.74	1.69	1.83	1.86	1.76	1.97
35	1.26	1.22	1.32	1.34	1.27	1.42

Table 5.2 Values of $(R_1 + R_n)/m$ for aluminium conductors at their normal operating temperature in milliohms/metre.

Conductor cross-sectional area mm^2	Insulation material	
	708C pvc	908C XLPE
1	2	3
16	4.58	4.89
25	2.88	3.07
35	2.08	2.22

For other assumed temperatures $(R_l + R_n)/m$ can be directly obtained from the mV/A/m values given in the volt drop sections of the tables in Appendix 4 of BS 7671.

$$(R_l + R_n)/m = \text{tabulated mV/A/m} \times \left(\frac{230 + t_g}{230 + t_p}\right) \text{ milliohms/m}$$

where t_g°C is the assumed temperature under short circuit conditions and t_p°C, as before, is the maximum permitted normal operating temperature for the type of insulation concerned.

In the first series of examples it is assumed that the overcurrent protective device has a breaking capacity greater than the prospective short circuit current at the point at which it is installed. Where HBC fuses such as those in BS 88 Pt 2 are used their high breaking capacity is such that this assumption is fully justified in the majority of installations and there is no need to invoke the second paragraph of Regulation 434–03–01. That paragraph allows the breaking capacity to be lower than the prospective short circuit current provided there is another protective device upstream having the necessary breaking capacity and the characteristics of the devices are co-ordinated so that the energy let-through of the devices can be withstood by the circuit conductors.

As shown in Figure 5.1 the calculated phase-to-neutral impedance at the remote end of the circuit has to be such that the short circuit current is greater than that corresponding to the point of intersection of the adiabatic line of the conductor and the time/current characteristic of the fuse. This short circuit current with the

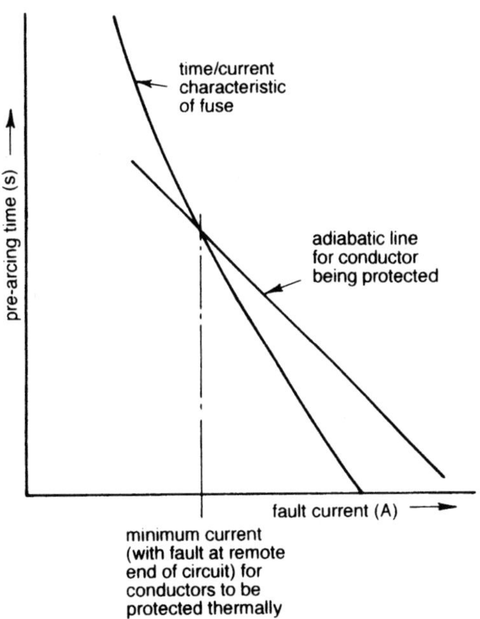

Figure 5.1 Fuse time/current characteristic showing minimum fault current for thermal protection of conductors having the adiabatic line shown.

fault at the remote end of the circuit is the minimum that can occur but Regulation 533–03–01 requires that account shall be taken of both minimum and maximum short circuit conditions.

As the position of the short circuit approaches the origin of the circuit so the short circuit current becomes increasingly greater and reaches its maximum value at the outgoing terminals of the fuse. Its maximum value is therefore the prospective fault current at the point of installation of the fuse and is determined by the phase-to-neutral impedance of the conductors upstream.

Should the short circuit occur close to the outgoing terminal of the fuse the disconnection time might well be less than 0.1 s and it then becomes necessary to check that the k^2S^2 for the conductors is greater than the let-through energy (I^2t) of the fuse, obtained from the I^2t characteristic provided by the fuse manufacturer. Figure 5.2 shows a typical example.

Some manufacturers provide this data in tabular form. In practice, when using HBC fuses it will generally be unnecessary to carry out this latter check but in the following examples some indication is given of the I^2t values encountered.

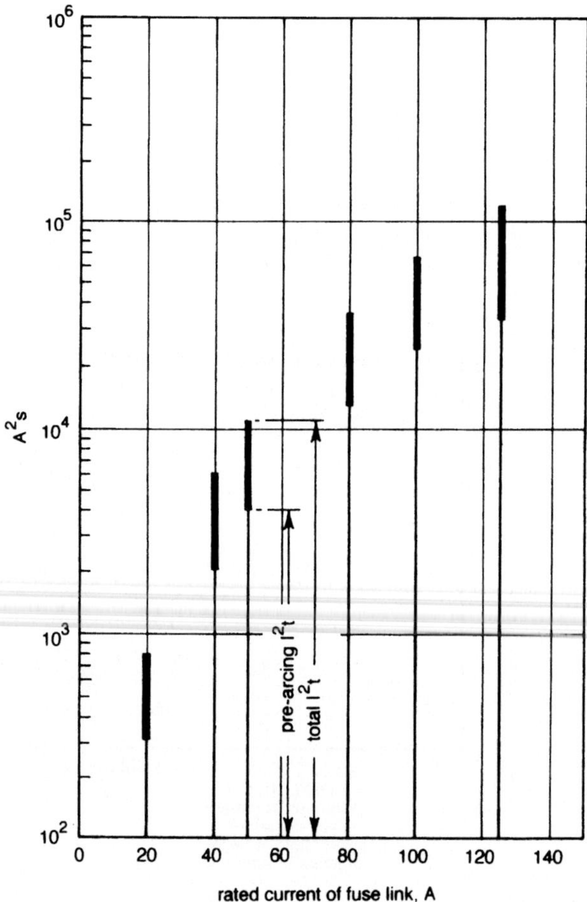

Figure 5.2 Typical I^2t characteristics for BS 88 'gG' fuses.

Example 5.1

A 230 V single-phase circuit is run in two-core (with cpc) 70°C pvc-insulated and sheathed cable to BS 6004 having copper conductors of 16 mm^2 cross-sectional area. It is protected only against short circuit by an 80 A BS 88 'gG' fuse.

If $l = 90$ m and the prospective short circuit current at the point of installation of the fuse is 3000 A check that the circuit complies with Regulation 434–03–03.

Answer

From Table 5.1 Column 2, for the circuit itself:

$$(R_l + R_n)/m = 2.76 \text{ milliohms/m}$$

$$(R_l + R_n) = \frac{2.76 \times 90}{1000} \text{ ohm} = 0.248 \text{ ohm}$$

Z_{pn} at the point of installation of the fuse is given by

$$\frac{230}{3000} \text{ ohm} = 0.077 \text{ ohm}$$

Z_{pn} at the remote end of the circuit is given by

$$(0.077 + 0.248) \text{ ohm} = 0.325 \text{ ohm}$$

$$I_{sc} = \frac{230}{0.325} \text{ A} = 708 \text{ A}$$

From the time/current characteristic of the fuse, as given in Appendix 3 of BS 7671, the disconnection time is 0.5 s.

From Table 43A, k = 115.

$$k^2S^2 = 115^2 \times 16^2 \text{ A}^2\text{s} = 3\,385\,600 \text{ A}^2\text{s}$$

$$I_{sc}^2t = 708^2 \times 0.5 \text{ A}^2\text{s} = 250\,632 \text{ A}^2\text{s}$$

Therefore Regulation 434–03–03 has been met, i.e. the circuit conductors are adequately protected against short circuit.

At the origin of the circuit, i.e. the point of installation of the overcurrent protective device, the short circuit current has been given as 3000 A. Examination of the time/current characteristic for the fuse shows that with this value of current the disconnection time is less than 0.01 s. Examination of the I^2t characteristic for the fuse indicates that its let-through energy is in the order of 30 000 A^2s – far less than the calculated k^2S^2 value for the conductors themselves. Thus even when the short circuit occurs close to the circuit origin the circuit conductors are adequately protected.

Example 5.2

A single-phase circuit is fed from the 110 V centre-tapped earth secondary of a transformer. The circuit is protected only against short circuit by a 63 A BS 88 'gG' fuse and is run in multicore non-armoured cable having 85°C rubber insulation and copper conductors clipped direct and not grouped with cables of other circuits. The live conductors are of $10\,mm^2$ cross-sectional area and $l = 25\,m$.

If the internal impedance of the transformer is 0.10 ohm check that the circuit complies with the adiabatic equation of Regulation 434–03–03.

Answer

From Table 5.1 Column 4, for the circuit itself:

$$(R_l + R_n)/m = 4.62\,\text{milliohms/m}$$

$$(R_l + R_n) = \frac{(4.62 \times 25)}{1000}\,\text{ohm} = 0.116\,\text{ohm}$$

Z_{pn} at the remote end of the circuit is given by:

$$(0.10 + 0.116)\,\text{ohm} = 0.216\,\text{ohm.}$$

$$I_{sc} = \frac{110}{0.216}\,A = 509\,A$$

Note that, whilst the terms $(R_l + R_n)$ and Z_{pn} have been used here, there is no 'neutral' as such.

From the time/current characteristic of the fuse, as given in Appendix 3 of BS 7671 the disconnection time is 0.4 s.

From Table 43A, $k = 134$.

$$k^2S^2 = 134^2 \times 10^2\,A^2s = 1\,795\,600\,A^2s$$

$$I_{sc}^2t = 509^2 \times 0.4\,A^2s = 103\,632\,A^2s$$

Therefore Regulation 434–03–03 is met.

As the position of the short circuit approaches the transformer terminals the fault current approaches

$$\frac{110}{0.10}\,A = 1100\,A$$

Examination of the I^2t characteristic for the fuse would show the circuit conductors were still adequately protected thermally.

Example 5.3

A 230 V single-phase circuit is run from a sub-distribution board at which the prospective short circuit current is 2500 A. The cables used are multi-core armoured 70°C pvc-insulated having copper conductors of

25 mm^2 cross-sectional area. The circuit is to be protected against short circuit by a 100 A BS 88 'gG' fuse.

If l = 47m check the circuit complies with Regulation 434–03–03.

Answer

From Table 5.1 Column 2, for the circuit itself:

$$(R_l + R_n)/m = 1.74\,\text{milliohms/m}$$

$$(R_l + R_n) = \frac{1.74 \times 47}{1000}\,\text{ohm} = 0.082\,\text{ohm}$$

Z_{pn} at the sub-distribution board is given by:

$$\frac{230}{2500}\,\text{ohm} = 0.092\,\text{ohm}$$

Z_{pn} at the remote end of the circuit is given by:

$$(0.082 + 0.092)\,\text{ohm} = 0.174\,\text{ohm}$$

$$I_{sc} = \frac{230}{0.174}\,\text{A} = 1322\,\text{A}$$

From the time/current characteristic of the fuse, as given in Appendix 3 of BS 7671, the disconnection time is 0.1 s.

From Table 43A, k = 115.

$$k^2S^2 = 115^2 \times 25^2\,\text{A}^2\text{s} = 8\,265\,625\,\text{A}^2\text{s}$$

$$I_{sc}^2t = 1322^2 \times 0.1\,\text{A}^2\text{s} = 174\,768\,\text{A}^2\text{s}$$

Therefore Regulation 434–03–03 is met.

One further example is given to show how one may use the mV/A/m values instead of resistance and reactance values.

Example 5.4

A 230 V single-phase circuit is run in a multicore non-armoured 70°C pvc-insulated cable having aluminium conductors of 70 mm^2 cross-sectional area and l = 55 m. The circuit is protected against short circuit only by an 80 A BS 88 'gG' fuse. It is fed from a sub-distribution board at which the calculated $(R_B + R_N)$ is 0.08 ohm and the calculated $(X_B + X_N)$ is 0.17 ohm.

Check that the circuit complies with Regulation 434–03–03.

Answer

From Table 4K2B Column 3, $(\text{mV/A/m})_r$ is 1.05 milliohms/m and $(\text{mV/A/m})_x$ is 0.165 milliohms/m. Thus $(R_l + R_n)/m$, being numerically the same as $(\text{mV/A/m})_r$ is 1.05 milliohms/m and $(X_l + X_n)/m$, being numerically the same as $(\text{mV/A/m})_x$ is 0.165 milliohms/m.

Thus:

$$(R_l + R_n) \text{ for the circuit itself} = \frac{1.05 \times 55}{1000} \text{ ohm} = 0.058 \text{ ohm}$$

$$(X_l + X_n) \text{ for the circuit itself} = \frac{0.165 \times 55}{1000} \text{ ohm} = 0.009 \text{ ohm}$$

Z_{pn} at the remote end of the circuit is given by:

$$Z_{pn} = \sqrt{(0.058 + 0.08)^2 + (0.009 + 0.17)^2} \text{ ohm} = 0.226 \text{ ohm}$$

So that

$$I_{sc} = \frac{230}{0.226} \text{ A} = 1018 \text{ A}$$

From the time/current characteristic of the fuse the disconnection time is 0.1 s. k = 76 (from Table 43A).

$$k^2S^2 = 76^2 \times 70^2 \text{ A}^2\text{s} = 28\,302\,400 \text{ A}^2\text{s}$$

$$I_{sc}^2 t = 1018^2 \times 0.1 \text{ A}^2\text{s} = 103\,632 \text{ A}^2\text{s}$$

Therefore Regulation 434–03–03 is met.

THE MORE RIGOROUS METHOD FOR A.C. SINGLE-PHASE CIRCUITS

As stated at the beginning of this chapter the second option open to the designer is to take account of the fact that the circuit conductors are probably not operating at the maximum permitted normal operating temperature for the type of insulation concerned. If this course of action is followed the conductor resistances under fault conditions will be adjusted as will be the value of k.

The method is exactly the same as the more rigorous method used for calculating earth fault loop impedances. Table 3.3 is used as the basis for the method remembering this gives the resistance per metre for one conductor at 20°C. The relevant value has then to be multiplied by the factor appropriate to the type of cable insulation from the following:

	Factor	
	Protective device listed in Appendix 3 of BS 7671	Protective device *not* listed in Appendix 3 of BS 7671
70°C pvc	$0.92 + 0.004t_1$	$1.24 + 0.002t_1$
85°C rubber	$0.92 + 0.004t_1$	$1.36 + 0.002t_1$
60°C rubber	$0.92 + 0.004t_1$	$1.32 + 0.002t_1$
90°C XLPE	$0.92 + 0.004t_1$	$1.42 + 0.002t_1$

The 'adjusted' k values are as given in Table 4.2 using the calculated operating temperature t_1 instead of t_i.

Example 5.5

A 230 V single-phase circuit is run in multicore armoured cable having XLPE insulation and copper conductors of 16 mm^2 cross-sectional area, clipped direct. The circuit is protected against short circuit only by a 100 A BS 88 'gG' fuse.

If Z_{pn} at the origin of the circuit is previously calculated to be 0.058 ohm, $I_b = 80$ A, $t_a = t_r = 30$°C and $l = 70$ m check compliance with Regulation 434–03–03.

Answer

First calculate t_l°C.

From Table 4E4A. Column 2, $I_{ta} = 110$ A.

$$t_l = 30 + \frac{80^2}{110^2}(90 - 30)°C = 61.7°C$$

From Table 3.3 and using the factor given above:

$$(R_l + R_n) = \frac{2 \times 1.15 \times [0.92 + (0.004 \times 61.7)] \times 70}{1000} \text{ ohm} = 0.188 \text{ ohm}$$

$$Z_{pn} = (0.188 + 0.058) \text{ ohm} = 0.246 \text{ ohm}$$

Thus:

$$I_{sc} = \frac{230}{0.246} \text{ A} = 935 \text{ A}$$

From the time/current characteristic given in Appendix 3, t = 0.4 s.

The k value from Table 4.2 is given by:

$$k = 192 - (0.55 \times 61.7) = 158$$

Thus:

$$k^2S^2 = 158^2 \times 16^2 \text{ A}^2\text{s} = 6\,390\,784 \text{ A}^2\text{s}$$

$$I_{sc}^2t = 935^2 \times 0.4 \text{ A}^2\text{s} = 349\,690 \text{ A}^2\text{s}$$

The circuit therefore complies with Regulation 434–03–03.

Close to the origin of the circuit the short circuit current approaches

$$\frac{230}{0.058} \text{ A} = 3966 \text{ A}$$

Checking with the manufacturer's I^2t characteristic would show if the circuit still complied.

Example 5.6

A single-phase 230 V circuit is fed from a transformer having an internal resistance of 0.06 ohm and an internal reactance of 0.11 ohm. The circuit is run in single-core 70°C pvc-insulated non-armoured cables with copper conductors of 50 mm^2 cross-sectional area installed on a perforated cable tray, not grouped with the cables of other circuits.

If $t_a = 10°C$, $I_b = 150$ A, $l = 130$ m and the overcurrent protective device providing short circuit protection only is a 160 A BS 88 'gG' fuse, check that the circuit complies with Regulation 434–03–03.

Answer

From Table 4D1A Column 8, $I_{ta} = 191$ A

$$t_l = 10 + \frac{150^2}{191^2}(70 - 30)°C = 34.7°C$$

From Table 4D1B Column 4, $(mV/A/m)_r = 0.93$ milliohms/m and $(mV/A/m)_x = 0.19$ milliohms/m. So that:

$$(R_l + R_n) = \frac{0.93}{1000} \times \left(\frac{230 + 34.7}{230 + 70}\right) \times 130 \text{ ohm} = 0.1067 \text{ ohm}$$

$$(X_l + X_n) = \frac{0.19 \times 130}{1000} \text{ ohm} = 0.0247 \text{ ohm}$$

$$Z_{pn} = \sqrt{(0.1067 + 0.06)^2 + (0.0247 + 0.11)^2} \text{ ohm} = 0.214 \text{ ohm}$$

Thus:

$$I_{sc} = \frac{230}{0.214} \text{ A} = 1075 \text{ A}$$

From the time/current characteristic given in Appendix 3, $t = 3$ s (approximately).

The k value from Table 4.2 is given by

$$k = 164 - (0.7 \times 34.7) = 139.7$$

$$k^2S^2 = 139.7^2 \times 50^2 \text{ A}^2\text{s} = 48\,790\,225 \text{ A}^2\text{s}$$

$$I_{sc}^2t = 1075^2 \times 3 \text{ A}^2\text{s} = 3\,466\,875 \text{ A}^2\text{s}$$

The circuit complies with Regulation 434–03–03.

Example 5. 7

A 230 V single-phase circuit from a sub-distribution board is run in single-core 70°C pvc-insulated cable in conduit. The cross-sectional area of the copper conductors is 16 mm^2. Z_{pn} at the sub-distribution board has been previously calculated to be 0.09 ohm. The circuit is protected by a 63 A BS 88 'gG' fuse.

If Z_{pn} at the remote end of the circuit is calculated to be 0.22 ohm check that the circuit complies with Regulation 434–03–03.

Answer

$$I_{sc} = \frac{230}{0.22}\ A = 1045\ A$$

From the time/current characteristic in Appendix 3 of BS 7671 the disconnection time is found to be less than 0.1 s. It is therefore necessary to obtain the I^2t characteristic of the fuse from the manufacturer. Let it be assumed it is found from this that the total $I^2t = 11300\ A^2s$. $k = 115$ (from Table 43A).

$$k^2S^2 = 115^2 \times 16^2\ A^2s = 3\,385\,600\ A^2s$$

As $k^2S^2 > I_{sc}^2t$ the circuit complies with Regulation 434–03–03.

It will be noted from this example that because the disconnection time is less than 0.10 s irrespective of the position of the fault, the short circuit current at the point of installation of the fuse being 2667 A, the I^2t characteristic and not the time/current characteristic has to be consulted.

When dealing with d.c. circuits the design method is exactly the same as that just described for single-phase a.c. circuits except that one then has internal resistances of the source and, irrespective of the cross-sectional areas of the conductors under consideration, one uses their resistances only, as reactances are no longer present.

All the examples in this chapter so far have concerned fuses. When the overcurrent protective device is a circuit breaker the design approach is a little different.

Figure 5.3 shows the time/current characteristic for a circuit breaker on which is superimposed the adiabatic line for the conductor being protected. The prospective short circuit current at the origin of the circuit should not exceed I_{max} because if it

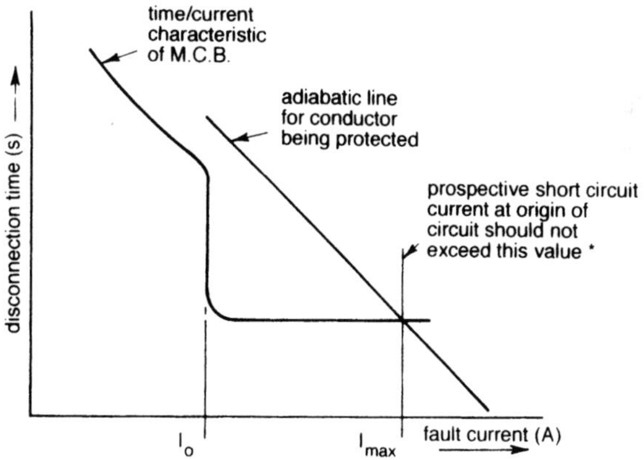
time/current
characteristic
of M.C.B.
adiabatic line
for conductor
being protected
prospective short circuit
current at origin of
circuit should not
exceed this value *
disconnection time (s)
I_o
I_{max}
fault current (A)
* NOTE:
If the prospective short circuit current at the origin
of the circuit does exceed I_{max} reference should
be made to the I^2t characteristic of the M.C.B.
I_{max} is also the minimum breaking capacity of the
M.C.B. unless the second paragraph of Regulation
434–03–01 is invoked

Figure 5.3 Relationship between circuit breaker time/current charcteristic and conductor adiabatic line showing maximum tolerable prospective short circuit current.

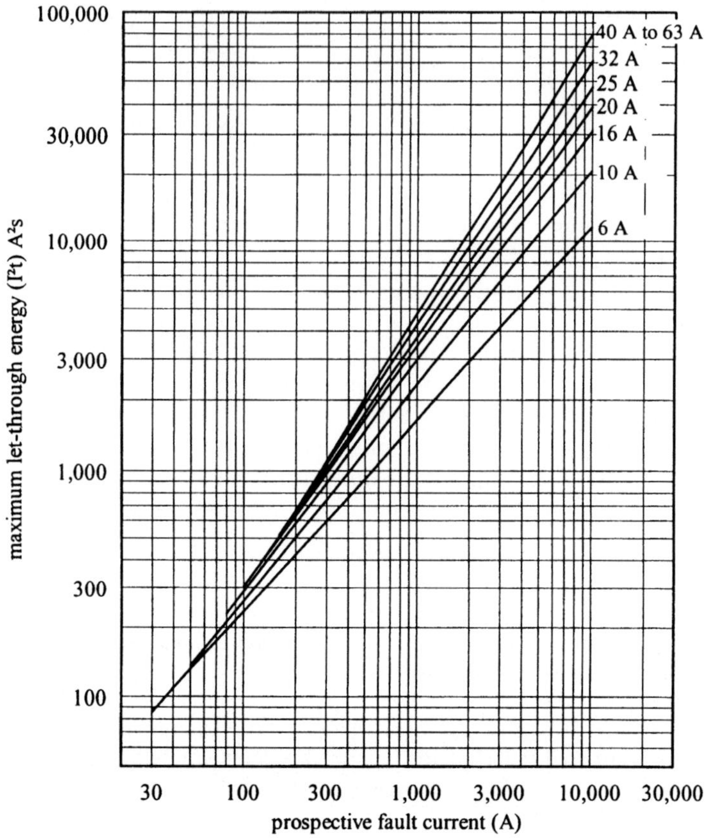

Figure 5.4 Typical let-through energy characteristics for Type B mcbs with a breaking capacity of 10 000 A.

did the circuit conductors would not be protected thermally over the whole length of the circuit.

Some manufacturers, in providing time/current characteristics, do not show the definite minimum time but do provide I^2t characteristics, a typical example being shown in Figure 5.4.

With current limiting mcbs, although I_{max} of Figure 5.3 might be exceeded, examination of the relevant I^2t characteristic might well show that the circuit conductors are, in fact, thermally protected along the whole length of the circuit.

Example 5.8

A 230 V single-phase circuit is run in a sheathed non-armoured cable having copper conductors and 85°C rubber insulation. The conductor cross-sectional area is 6 mm^2. The circuit is taken from a sub-distribution board at which Z_{pn} is 0.028 ohm.

If the circuit length is 35 m and it is intended to give overload and short circuit protection by using a 40 A Type B mcb, determine the minimum breaking capacity of the mcb and check compliance with Regulation 434–03–03.

Answer

The short circuit current at the sub-distribution board is given by:

$$\frac{230}{0.028}\,\mathrm{A} = 8214\,\mathrm{A}$$

The breaking capacity for the Type B mcb, selected from the standard values given in BS EN 60898, must be 10 000 A.

Despite complying with Regulation 434-03-02 it is worthwhile checking for compliance with Regulation 434-03-03. Because the fault current at the sub-distribution board is greater than that which will provide instantaneous operation of the mcb, according to the time/current characteristics given in Appendix 3 of BS 7671, manufacturer's I^2t data should be consulted. Typical data for a 40 A mcb indicates that the let-through energy at a prospective short-circuit current of approximately 8000 A is about 60 000 A^2s. In the present case $k^2S^2 = 134^2 \times 6^2\ A^2s = 646\,416\ A^2s$, so if this particular make of mcb is used, the circuit is fully protected over its whole length and Regulation 434-03-03 is met.

It is essential that where this approach is used the I^2t characteristic must be that from the manufacturer of the mcb it is intended to use, not those of another manufacturer.

There is a need to check the current when the short circuit occurs at the remote end of the circuit.

From Table 5.1:

$$(R_l + R_n)/m = 7.76\,\text{milliohms/m}$$

$$(R_l + R_n) = \frac{7.76 \times 35}{1000}\,\text{ohm} = 0.27\,\text{ohm}$$

$$I_{sc} = \frac{230}{0.028 + 0.27}\,\mathrm{A} = 772\,\mathrm{A}$$

Again the prospective fault current is in excess of the maximum value given in Appendix 3 of BS 7671 for a 40 A Type B mcb and hence the manufacturer's let-through energy data should be consulted. However, it is clear from the previous calculation for the fault current at the sub-distribution board that the circuit is thermally protected.

Sometimes there is no need to carry out the last calculation. I_o in Figure 5.3 is also the minimum earth fault current to give protection against indirect contact and because of the impedances involved will in general be less than the short circuit current. Theoretically, however, if the installation is part of a TN–C–S system and mineral-insulated cable is used, the earth fault current will be greater than the short circuit current. It is therefore necessary to check that the latter is still sufficient to bring out the mcb within its so-called 'instantaneous' time, i.e. within 0.1 s.

A.C. THREE-PHASE CIRCUITS

For three-phase circuits the calculations are a little more involved compared with those for single-phase circuits because there are now three types of short circuit to consider.

(a) The symmetrical three-phase short circuit current is given by:

$$I_{sc} = \frac{U_p}{\sqrt{(R_B + R_l)^2 + (X_B + X_l)^2}} \text{ A}$$

Note in particular the absence of R_N, X_N, R_n and X_n from the above.

(b) The short circuit current phase-to-phase is given by:

$$I_{sc} = \frac{0.87\,U_p}{\sqrt{(R_B + R_l)^2 + (X_B + X_l)^2}} \text{ A}$$

Note again the absence of R_N, X_N, R_n and X_n.

(c) The short circuit current phase-to-neutral is given by:

$$I_{sc} = \frac{U_p}{\sqrt{(R_B + R_N + R_l + R_n)^2 + (X_B + X_N + X_l + X_n)^2}} \text{ A}$$

This is exactly the same as the expression for the single-phase case.
The above symbols have the same significance as before but R_B and X_B are for *one phase* of the supply and R_l and X_l are for *one of the phase conductors* of the protected circuit.

$$U_p = \frac{\text{line-to-line voltage}}{\sqrt{3}} \text{ V}$$

The symmetrical three-phase fault current is always greater than the phase-to-phase short circuit current which, in turn, is always greater than the phase-to-neutral short circuit current.

Thus, as illustrated by Figure 5.5, when checking if the live conductors of a three-phase circuit are adequately protected thermally when the protective device is a fuse i.e. that the circuit complies with the adiabatic equation, it is necessary to use the *minimum* of the three values. Thus if the neutral is not being distributed the short circuit current phase-to-phase has to be calculated and used in the adiabatic equation. If the neutral is distributed, the phase-to-phase value is not needed to be calculated and compliance with the adiabatic equation is checked using the phase-to-neutral short circuit current.

When making the necessary calculations the values of R_l and R_n are those at the assumed temperature under fault conditions taking account, if one wishes, of the fact that the initial temperature may well be less than the maximum normal operating temperature. The values of R_B and R_N are assumed not to increase because of the fault in the final circuit and are taken to be those at the maximum normal operating temperature for the cable insulation concerned (or if one wishes, at t_l°C the actual operating temperature under normal load conditions).

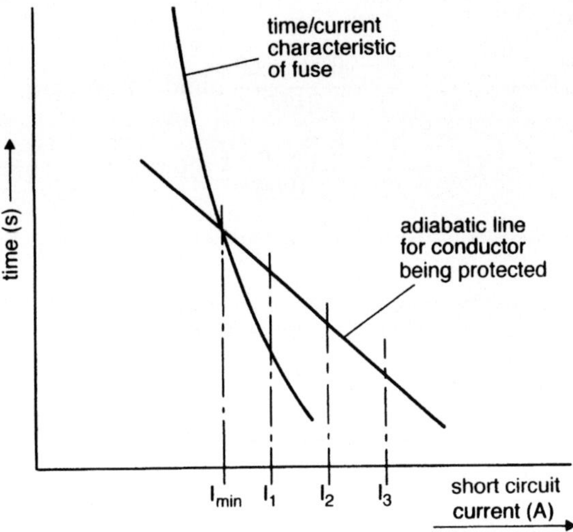

I_1 = phase–neutral short circuit current
I_2 = phase–phase short circuit current
I_3 = symmetrical three–phase short circuit current
I_{min} = the minimum short circuit current for compliance with Regulation 434–03–03

Figure 5.5 Relationship between fuse time/current characteristic, conductor adiabatic line and three-phase short circuit currents.

In the next three examples because short circuit protection is being provided by BS 88 'gG' fuses it is assumed their breaking capacity is adequate and it is not necessary to invoke the second paragraph of Regulation 434–03–01.

Example 5.9

A 400 V three-phase four-wire circuit is fed from a sub-distribution board. The previously calculated resistance of the source and distribution circuit per phase up to that board is 0.035 ohm and the reactance is 0.05 ohm. The circuit is run in a multicore armoured cable having XLPE insulation and copper conductors of 50 mm^2 cross-sectional area. The cable is clipped direct, $l = 60$ m and $t_a = 30°C$.

If the circuit is protected against short circuit only by a 200 A BS 88 'gG' fuse check the circuit complies with Regulation 434–03–03.

Answer

Because the neutral is distributed there is need only to determine the phase-to-neutral short circuit current.

From Table 4E4B Column 4:

$$(mV/A/m)_r = 0.86\,\text{milliohms/m}; \quad (mV/A/m)_x = 0.135\,\text{milliohms/m}.$$

Thus:

$$(R_l + R_n) = \frac{2}{\sqrt{3}} \times \frac{0.86 \times 60}{1000} \text{ ohm} = 0.06 \text{ ohm}$$

$$(X_l + X_n) = \frac{2}{\sqrt{3}} \times \frac{0.135 \times 60}{1000} \text{ ohm} = 0.0094 \text{ ohm}$$

The phase-to-neutral short circuit current is given by:

$$\frac{230}{\sqrt{(0.035 + 0.06)^2 + (0.05 + 0.0094)^2}} \text{ A} = 2053 \text{ A}$$

From Table 43A, $k = 143$.

From the time/current characteristic in Appendix 3 of BS 7671 the disconnection time is found to be 0.5 s.

$$k^2S^2 = 143^2 \times 50^2 \text{ A}^2\text{s} = 51\,122\,500 \text{ A}^2\text{s}$$

$$I_{sc}^2 t = 2053^2 \times 0.5 \text{ A}^2\text{s} = 2\,107\,404 \text{ A}^2\text{s}$$

The circuit therefore complies with Regulation 434–03–03.

Example 5.10

A 400 V three-phase circuit is fed from a sub-distribution board where the prospective short circuit current has already been estimated to be 14 000 A. The circuit is run in multicore armoured 70°C pvc-insulated and sheathed cables having copper conductors of 10 mm^2 cross-sectional area.

If $l = 62$ m and the circuit is protected against short circuit by 100 A BS 88 'gG' fuses check compliance with Regulation 434–03–03. The neutral is not distributed.

Answer

Because the neutral is not distributed the phase-to-phase short circuit current with the fault occurring at the remote end of the circuit has to be calculated.

$$Z_D = \frac{230}{14\,000} \text{ ohm} = 0.016 \text{ ohm}$$

From Table 5.1 Column 2:

$$R_l/m = \frac{(R_l + R_n)/m}{2} = \frac{4.39}{2} \text{ milliohms/m} = 2.19 \text{ milliohms/m}$$

$$R_l = \frac{2.19 \times 62}{1000} \text{ ohm} = 0.136 \text{ ohm}$$

The short circuit current phase-to-phase is given by:

$$I_{sc} = \frac{0.87 \times 230}{0.016 + 0.136} \text{ A} = 1316 \text{ A}$$

From the time/current characteristic given in Appendix 3 of BS 7671 the disconnection time is found to be approximately 0.1 s.

From Table 43A, k = 115.

$$k^2S^2 = 115^2 \times 10^2\ A^2s = 1\,322\,500\ A^2s$$

$$I_{sc}^2t = 1316^2 \times 0.1\ A^2s = 173\,186\ A^2s$$

The circuit therefore complies with Regulation 434–03–03.

A typical value for the energy let-through I^2t for a 100 A fuse will be in the order of 60 000 A^2s so there is no doubt that the circuit complies irrespective of the position of the fault along its length.

Example 5.11

A 400 V three-phase circuit is fed from a distribution board where R_B has been previously calculated to be 0.02 ohm and X_B to be 0.06 ohm. The circuit is run in single-core armoured 70°C pvc-insulated cables having aluminium conductors of 50 mm^2 cross-sectional area (in trefoil touching) protected against short circuit by 160 A BS 88 'gG' fuses.

If the circuit length is 85 m and the neutral is not distributed check for compliance with Regulation 434–03–03.

Answer

From Table 4K3B Column 5, $(mV/A/m)_r = 1.35$ milliohms/m and $(mV/A/m)_x = 0.195$ milliohms/m.

As the neutral is not distributed the phase-to-phase short circuit current with the fault at the remote end of the circuit has to be calculated.

$$R_l/m = \frac{1.35}{\sqrt{3}}\ \text{milliohms/m} = 0.78\ \text{milliohms/m}$$

$$R_l = \frac{0.78 \times 85}{1000}\ \text{ohm} = 0.066\ \text{ohm}$$

$$X_l/m = \frac{0.195}{\sqrt{3}}\ \text{milliohms/m} = 0.113\ \text{milliohms/m}$$

$$X_l = \frac{0.113 \times 85}{1000}\ \text{ohm} = 0.01\ \text{ohm}$$

The short circuit current phase-to-phase is given by:

$$\frac{0.87 \times 230}{\sqrt{(0.02 + 0.066)^2 + (0.06 + 0.01)^2}}\ A = 1805\ A$$

From the time/current characteristic in Appendix 3 of BS 7671 the disconnection time is approximately 0.3 s.

From Table 43A, k = 76.

$$k^2S^2 = 76^2 \times 50^2\ A^2s = 14\,440\,000\ A^2s$$

$$I_{sc}^2t = 1805^2 \times 0.3\ A^2s = 977\,407\ A^2s$$

The circuit therefore complies with Regulation 434–03–03.

The symmetrical three-phase currents are given by:

(i) at the point of installation of the fuses:

$$\frac{230}{\sqrt{0.02^2 + 0.06^2}}\ A = 3637\ A$$

(ii) at the remote end of the circuit:

$$\frac{1805}{0.87}\ A = 2075\ A$$

and it will be found the circuit still complies.

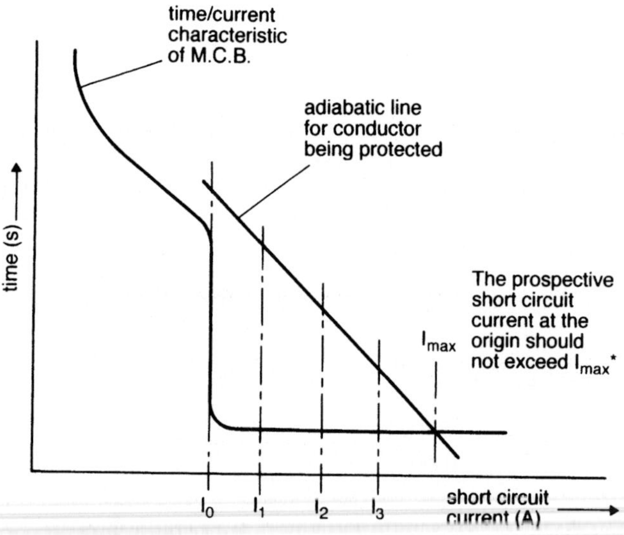

I_0 = minimum current for 0.1s disconnection
I_1 = phase–neutral short circuit current
I_2 = phase–phase short circuit current
I_3 = symmetrical three–phase short circuit current

*NOTE: If the prospective short circuit current at the origin of the circuit does exceed I_{max} reference should be made to the I^2t characteristic of the M.C.B. I_{max} is also the minmum breaking capacity of M.C.Bs unless the second paragraph of Regulation 434–03–01 is invoked

Figure 5.6 Relationship between mcb time/current characteristic, conductor adiabatic line and three-phase short circuit currents.

Figure 5.6 shows the time/current characteristics for an mcb on which is superimposed the adiabatic line for the conductor being protected and the various short circuit currents that occur in three-phase circuits are also shown. All the previous remarks concerning single-phase circuits protected by mcbs still apply.

As shown in Figure 5.6 the three types of short circuit current should be between I_o and I_{max}. I_{max} is again the prospective short circuit current at the origin of the circuit and is also the minimum breaking capacity for the mcb unless the second paragraph of Regulation 434–03–01 is invoked, but now I_{max} is related to the symmetrical three-phase short circuit condition.

Whilst I_{max} has been described as the current above which the circuit conductors will not be protected thermally it must be reiterated that with current-limiting mcbs one should refer to the I^2t characteristics for the device and it will be found that even with currents in excess of I_{max} compliance with Regulation 434–03–03 may well be obtained.

Example 5.12

A 400 V three-phase four-wire circuit is run in a multicore 70°C pvc-insulated and sheathed cable having copper conductors of 16 mm^2 cross-sectional area installed in trunking with other cables. The circuit is fed from a sub-distribution board where the previously calculated value of R_B is 0.045 ohm and of X_B is 0.02 ohm. The previously calculated value of R_N is 0.015 ohm and of X_N is 0.01 ohm.

If the length of the circuit is 90 m and it is to be protected against overload and short circuit by 50 A Type B mcbs, determine the minimum breaking capacity of the mcb and check compliance with Regulation 434–03–03.

Answer

The symmetrical three-phase short circuit current at the origin of the circuit is given by:

$$\frac{230}{\sqrt{0.045^2 + 0.02^2}} \text{ A} = 4671 \text{ A}$$

The breaking capacity of the mcb should be greater than 4671 A. The selection of a mcb with a breaking capacity of at least 6000 A is therefore acceptable.

Assume that from the manufacturer's I^2t characteristics it is found that the let-through energy for a 50 A mcb at the prospective short circuit current is 32 000 A^{2}s. k = 115, so that

$$k^2S^2 = 115^2 \times 16^2 \text{ A}^2\text{s} = 3\,385\,600 \text{ A}^2\text{s}$$

The conductors are thermally protected.

(According to Regulation 434–03–02 there is, in fact, no need to carry out this check.)

As the neutral is distributed now check that the phase-to-neutral short circuit current with the fault at the remote end of the circuit is sufficient to give disconnection in 0.1 s.

From Table 5.1 Column 2:

$$(R_l + R_n)/m = 2.76\,\text{milliohms/m}$$

$$R_l + R_n = \frac{2.76 \times 90}{1000}\,\text{ohm} = 0.248\,\text{ohm}$$

The phase-to-neutral short circuit current is given by:

$$\frac{230}{\sqrt{(0.045 + 0.015 + 0.248)^2 + (0.02 + 0.01)^2}}\,\text{A} = 743\,\text{A}$$

This prospective fault current is in excess of the maximum value given in Appendix 3 of BS 7671 for a 50 A Type B mcb and hence the manufacturer's let-through energy data should be consulted.

If the manufacturer's data indicates that the let-through energy of a 50 A mcb, for a fault current of about 750 A, is 3300 A^2s, then this is less than k^2S^2 for the cable, as calculated above. Thus the conductors are protected thermally for all short circuit conditions along the whole length of the circuit.

It is not the purpose of this book to make comparisons between fuses and mcbs (or mccbs) but in the present context it has to be said that one disadvantage of mcbs is that their breaking capacities are significantly less than those of HBC fuses to BS 88 Pt 2. Thus, cases arise when mcbs are to be installed where the prospective short circuit current is greater than their breaking capacity and it becomes necessary for the designer to invoke the second paragraph of Regulation 434–03–01. The most common example of this is the consumer unit using mcbs.

The Electricity Association has stated in Engineering Recommendation P25 that the maximum design value of the prospective short circuit current at the point of connection of the service cable to an Electricity Supply Board's main low voltage distribution should be taken as 16 kA. In areas other than those served by London Electricity attenuation by the service cable can be allowed for in estimating the prospective short circuit current at the origin of the installations. Electricity companies have provided data to enable the attenuation in the service cable to be calculated.

Because many mcbs have breaking capacities less than 16 kA it was decided to introduce in the British Standard for consumer units the concept of a conditional rating for the combination of the mcb concerned and the back-up protection afforded by the electricity company's 100 A cut-out fuse.

More generally, manufacturers are now producing data which enables the designer to correctly co-ordinate the characteristics of devices where the second paragraph of Regulation 434–03–01 has to be invoked, one example being shown in Figure 5.7.

This shows that for a 32 A mcb of one particular manufacturer, which has a breaking capacity of 10000 A and is in accordance with BS EN 60898, this may be installed at a point where the prospective short circuit current is greater than 10 000 A. This is only acceptable provided that upstream there is a BS 88 'gG' fuse having a current rating of up to 160 A.

Some indication of the increase of prospective short circuit current over the rated breaking capacity of moulded case circuit breakers to BS EN 60947-2 that can be tolerated where back-up protection is provided can be obtained by first determining the cut-off current of the fuse providing that protection. This cut-off current is obtained from the manufacturer's characteristics, an example being given in Figure 5.8 from which it is seen that the cut-off currents in *peak* amperes are plotted against prospective current in rms symmetrical amperes.

Having calculated the prospective short circuit current at the point of installation *of the mcb* the corresponding peak cut-off current is obtained from the cut-off characteristic of the fuse.

BS EN 600947-2 gives factors of minimum short circuit making capacity to rated short circuit breaking capacity as in Table 5.3.

For proper co-ordination to be achieved the rated breaking capacity of the mccb has to be such that when multiplied by the factor given in the above table the resulting product has to be greater than the peak cut-off current of the fuse.

A similar approach can be employed for mcbs to BS EN 60898 although this standard makes no distinction between values of short circuit making capacity and short circuit breaking capacity.

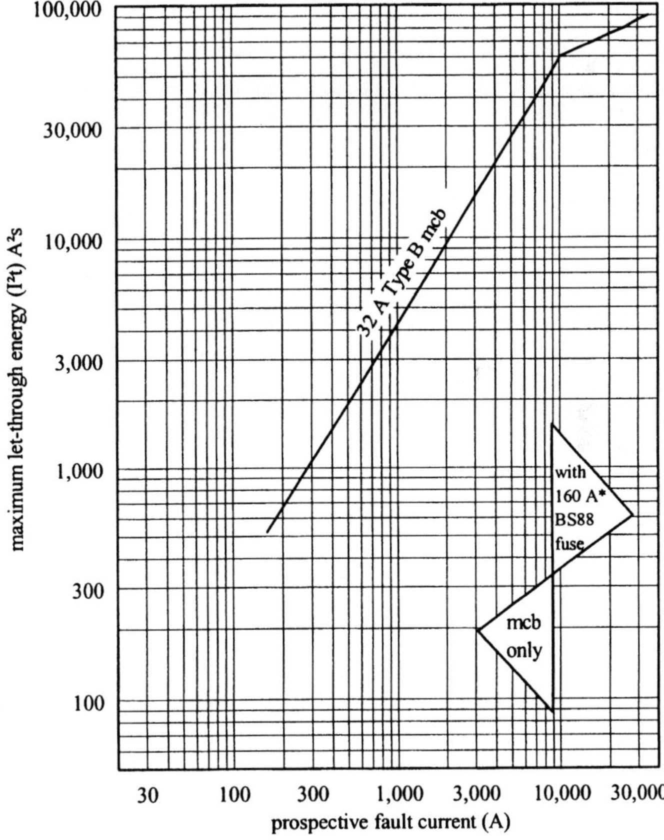

* I for the fuse may be less than 160 A

Figure 5.7 Typical let through energy characteristic for a Type B mcb with a breaking capacity of 10 000 A showing the effect of backup protection by BS88 'gG' fuse.

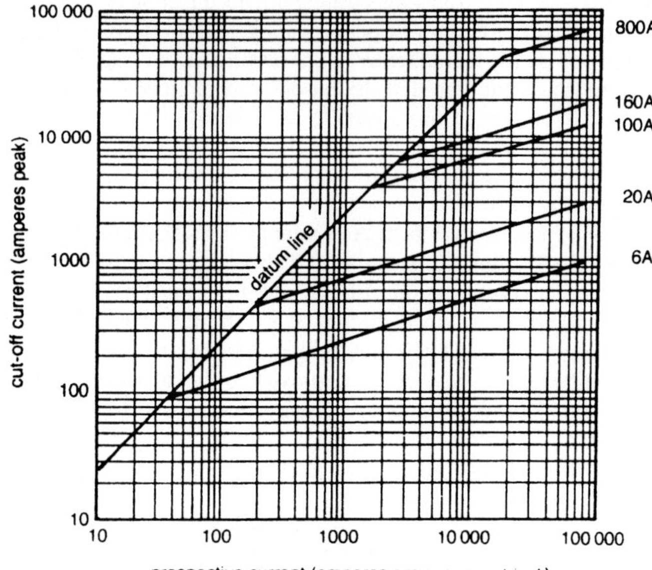

Figure 5.8 Typical cut-off current characteristic for BS 88 'gG' fuses.

Table 5.3 Factor for minimum making capacity to rated breaking capacity.

Rated breaking capacity (I_{cn}) ka	Factor
$4.5 < I_{cs} \leq 6$	1.5
$6 < I_{cs} \leq 10$	1.7
$10 < I_{cs} \leq 20$	2.0
$20 < I_{cs} \leq 50$	2.1
$50 < I_{cs}$	2.2

Example 5.13

A 400 V three-phase three-wire circuit is taken from a distribution board at which the prospective short circuit current (symmetrical three-phase) has been previously calculated to be 30 kA. The circuit is run in multicore armoured 70°C pvc-insulated cable having copper conductors of 4 mm^2 cross-sectional area, the cable being clipped direct and not grouped with other cables. l = 55 m. The circuit is protected against overload and short circuit by a 32 A Type B mcb having a breaking capacity of 10 000 A. The circuit to the distribution board is protected by 100 A BS 88 ‘gG’ fuses.

Check that the circuit complies with Regulation 434–03–01.

Answer

Because the prospective short circuit current at the point of installation is greater than the breaking capacity of the mcb it is necessary to obtain the necessary data from the manufacturer.

Assume that information is as given in Figure 5.7. From this it is seen that the maximum total let-through energy with the stated prospective short circuit current of 30 000 A is approximately 85 000 A^2s.

From Table 43A of BS 7671, $k = 115$. Thus:

$$k^2S^2 = 115^2 \times 4^2\ A^2s = 211\,600\ A^2s$$

As this is greater than the I^2t for the mcb plus fuse the circuit complies.

Because the neutral is not distributed check the phase-to-phase short circuit current when the fault is at the remote end of the circuit.

From Table 5.1 Column 2:

$$(R_1 + R_n)/m = 11.06\ \text{milliohms/m}$$

so that:

$$R_1/m = 5.53\ \text{milliohms/m}$$

$$R_1 = \frac{5.53 \times 55}{1000}\ \text{ohm} = 0.304\ \text{ohm}$$

Only Z_B is known and is given by:

$$Z_B = \frac{230}{30\,000}\ \text{ohm} = 0.0077\ \text{ohm}$$

The phase-to-phase short circuit current with the fault occurring at the far end is given by:

$$I_{sc} = \frac{0.87 \times 230}{0.0077 + 0.304}\ \text{A} = 642\ \text{A}$$

The time/current characteristics in Appendix 3 of BS 7671 indicate that, for this level of fault current, the let-through energy should be taken from the manufacturer’s data. If this data is as given in Figure 5.7, then the let-through energy would be approximately 2400 A^2s.

In all the examples so far there has been no need to calculate the relevant impedances from the source of energy to the origin of the circuit under consideration, because that information has been provided in some form or other.

When calculating the prospective short circuit current at the origin of a circuit this, of course, is determined by the impedances upstream of that origin and because one is calculating the worst case those impedances should be calculated at ambient temperature values because the fault could occur in a 'cold' installation, i.e. in one which has just begun to supply load current. For this purpose the ambient temperature is taken to be 20°C (some designers use a lesser value) and not the reference ambient temperature of 30°C associated with the tables of current-carrying capacities in Appendix 4 of BS 7671.

As the cross-sectional areas of cable conductors increase it will be found that the reactance component of the impedance becomes greater than the resistance component so any advantage gained in correcting the latter to the actual operating temperature decreases. For the largest sizes, particularly for single-core cables, the reactance component is considerably greater than the resistance component and it is quite common practice to use the former (which is independent of temperature) to calculate prospective fault currents.

When calculating the fault current at the remote end of a circuit for the purpose of checking that the circuit conductors are thermally protected the impedance values for the circuit itself are those at the assumed temperature under fault conditions and any circuit upstream of it is taken to be at the appropriate maximum normal operating temperature (t_p°C) or at the calculated actual operating temperature (t_1°C). This is shown diagrammatically in Figure 5.9, and in Chapter 6 an example is given.

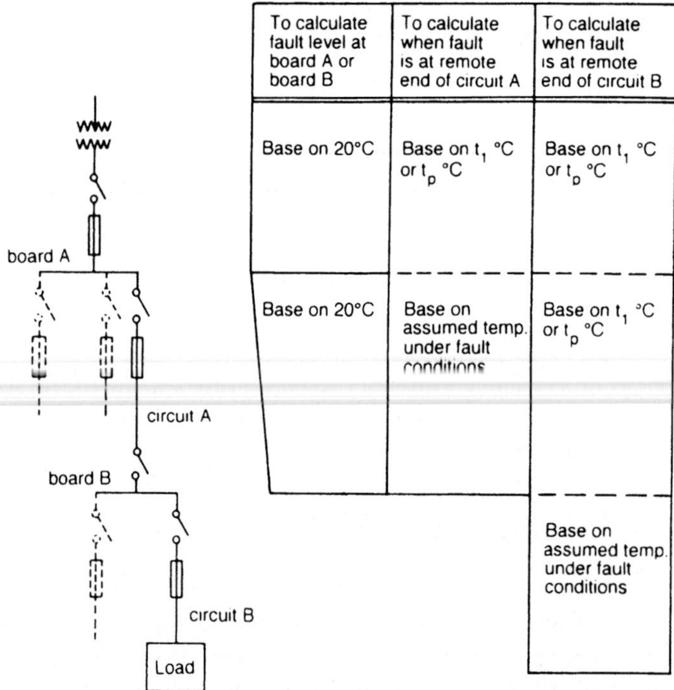

Figure 5.9 Temperatures at which conductor resistances should be calculated.

Chapter 6
Combined Examples

The previous chapters have dealt with individual aspects of the design of circuits:

Chapter 1 dealt with the calculation of conductor cross-sectional areas for compliance with Regulation 523–01–01: namely that under normal load conditions the maximum permitted normal operating temperature is not exceeded.

Chapter 2 dealt with the calculation of voltage drop which in very many cases is the limiting factor in design.

Chapter 3 dealt with the calculation of earth fault loop impedance for compliance with the requirement concerning maximum disconnection times of protective devices being used for protection against indirect contact.

Chapter 4 dealt with the calculations concerning protective conductor cross-sectional areas for compliance with Regulation 543–01–01.

Chapter 5 dealt with the calculations associated with short circuit conditions.

In each of these chapters a simple design method is given and, where appropriate, a more accurate or rigorous design method, together with some indication of the advantage that might be gained by the use of the latter. The purpose of this chapter, as its title indicates, is to combine all the calculations in order to completely design a circuit.

In undertaking the 'manual' design of any circuit one should *always* use the simple method. Only if the resulting design leads to non-compliance with the relevant requirements should consideration be given to redesigning using the more accurate or rigorous methods because these might avoid the need to increase conductor cross-sectional areas and hence the material costs.

Circumstances might arise where, in any event, it may be cheaper to increase conductor cross-sectional areas than to incur additional design costs. Each case has to be treated on its own merits and the experience of the designer should indicate the appropriate course of action to take.

Where computer programs are used there is no reason why these should not adopt some, if not all, of the more rigorous design methods.

Where it is intended to use the so-called standard circuit arrangements such as the 30 A ring circuit or the 20 A radial circuit the design requirement is normally limited to ensuring that the circuit lengths are such that the resulting impedances are within the specified maxima. Only if such circuits are grouped or are operating in a high ambient temperature are any circuit calculations required. For this reason the great majority of the examples in this chapter are of 'non-standard' circuit arrangements.

Having chosen the type of cable to use and the method of installation, e.g. whether grouped or run singly and whether clipped direct or enclosed, the route the circuit concerned is intended to take is determined and the circuit route length then estimated. That estimation should be as accurate as is reasonably practicable because this determines, to some extent, the accuracy of the calculated values of the resistances and impedances of the circuit conductors. This point is an important one because the calculation of conductor resistances and impedances is a major part of the design of circuits.

Ignoring for the present the assessment of the breaking capacities of protective devices, the calculations used in designing circuits are:

(a) those for determining the minimum conductor cross-sectional areas that can be used for compliance with Regulation 523–01–01 and, where appropriate, with the requirements for overload protection prescribed in Regulation 433–02–01;
(b) those for checking compliance with the voltage drop limitation prescribed in Regulation 525–01–01;
(c) those for determining conductor resistances or impedances in order to check compliance, where appropriate, with the requirements of Section 413 (protection against indirect contact using automatic disconnection of supply), Regulation 434–03–03 (short circuit protection) and Section 543 (protective conductors).

As regards item (c) the designer is faced with a particular dilemma because, as shown in Figures 3.1 and 5.9, he may have to calculate the circuit resistances and impedances at more than one temperature. Furthermore, whilst the designer for item (b) may use the tabulated mV/A/m values these are really resistance and impedance values at the maximum permitted normal operating temperature for the type of cable insulation concerned.

Having carried out all the necessary calculations the designer has to indicate to the person carrying out the tests to verify compliance with BS 7671 the values of resistance and impedance which should not be exceeded in those tests. Obviously the tester does not want to correct the values specified by the designer to the temperature at which the tests will be undertaken. The tester will expect that conversion to be done by the designer.

It is suggested that 20°C is a reasonable temperature to pick as the test temperature although, in practice, cases will arise where the actual ambient temperature may be considerably different and in such cases some account has to be taken of that fact. Even if a circuit has been designed say for a high ambient temperature it does not necessarily follow that will be the actual temperature during the tests for compliance. It is therefore not possible to lay down a hard and fast rule on this matter but for the purposes of illustration in some of the following examples the test values of conductor resistance or impedance are based on 20°C.

The first series of examples are for individual circuits where the assumption has been made that the breaking capacity of the associated protective device is adequate.

Example 6.1

A 230 V single-phase circuit is to be run in flat two-core (with cpc) 70°C pvc-insulated and sheathed cable having copper conductors clipped direct and not grouped with the cables of other circuits. The circuit is to be protected by a BS 88 'gG' fuse against overload and short circuit and this fuse is also to give protection against indirect contact with a maximum disconnection time of 5 s.

If $I_b = 42$ A, $l = 27$ m, $t_a = 40$°C and $Z_E = 0.8$ ohm, determine the nominal rating of the fuse and the minimum conductor cross-sectional area that can be used if the voltage drop is not to exceed 4% of the nominal voltage.

Answer

Because $I_n \geqslant I_b$ choose from the standard ratings of BS 88 'gG' fuses $I_n = 50$ A.

From Table 4C1, $C_a = 0.87$.

As $C_g = 1$ and $C_i = 1$

$$I_t = 50 \times \frac{1}{0.87} \text{ A} = 57.5 \text{ A}.$$

From Table 4D5A Column 4 it is found that the minimum conductor cross-sectional area that can be used is 10 mm^2 having $I_{ta} = 64$ A.

From Column 3 Table 4D2B the mV/A/m = 4.4 milliohms/m. Hence

$$\text{voltage drop} = \frac{4.4 \times 42 \times 27}{1000} \text{ V} = 5 \text{ V} = 2.17\% \text{ of } 230 \text{ V}$$

The circuit therefore meets the limitation in voltage drop.

The cross-sectional area of the protective conductor for this standard cable is 4 mm^2.

Table 54C applies and from Table 3.1 Column 4, $(R_1 + R_2)$/m is found to be 7.73 milliohms/m.

$$R_1 + R_2 = \frac{7.73 \times 27}{1000} \text{ ohm} = 0.21 \text{ ohm}$$

$$Z_s = (0.8 + 0.21) \text{ ohm} = 1.01 \text{ ohm}$$

The circuit therefore meets the maximum value of 1.09 ohm given in Table 41D.

It remains to check whether the cpc meets the adiabatic equation given in Regulation 543–01–03.

From Table 54C, k = 115.

The earth fault current I_{ef} is given by:

$$\frac{230}{1.01} \text{ A} = 228 \text{ A}$$

From the time/current characteristic given in Appendix 3 of BS 7671 the disconnection time is found to be approximately 4 s.

For compliance $k^2S^2 \geqslant I^2t$

$$k^2S^2 = 115^2 \times 4^2\ A^2s = 211\,600\ A^2s$$

$$I_{ef}^2t = 228^2 \times 4\ A^2s = 207\,936\ A^2s$$

The adiabatic equation is therefore met.

To obtain the *test* value for $(R_1 + R_2)$ use Table 3.3 from which it is found to be:

$$\frac{(1.83 + 4.61) \times 27}{1000}\ \text{ohm} = 0.174\ \text{ohm at } 20°\text{C}.$$

Having assumed that the BS 88 'gG' fuse has adequate breaking capacity Regulation 434–03–02 applies and there is no need to check that the circuit complies with the adiabatic equation under *short circuit* conditions, because that fuse is providing overload as well as short circuit protection.

As the circuit complies with all the requirements there has been no point in using other than the simple design approach. However, remembering that the mV/A/m values for single-phase circuits are numerically twice the milliohms/m values for a single conductor one could have used these to determine $(R_1 + R_2)$.

From Column 3 of Table 4D2B, mV/A/m for 4 mm^2 = 11 milliohms/m and mV/A/m for 10 mm^2 = 4.4 milliohms/m. Thus:

$$(R_1 + R_2)/\text{m at } 70°\text{C} = \frac{11 + 4.4}{2}\ \text{milliohms/m} = 7.7\ \text{milliohms/m}$$

$$(R_1 + R_2)/\text{m at } 20°\text{C} = \left(\frac{230 + 20}{230 + 70}\right) \times 7.7\ \text{milliohms/m}$$

$$= 6.42\ \text{milliohms/m}$$

Using these values one would obtain:

$$(R_1 + R_2)\ \text{at } 70°\text{C} = 0.208\ \text{ohm}$$

$$(R_1 + R_2)\ \text{at } 20°\text{C} = 0.173\ \text{ohm}$$

Example 6.2

A 230 V single-phase circuit is to be run in single-core 70°C pvc-insulated cables having copper conductors in trunking with five other similar circuits. The circuit is to be protected by a BS 88 'gG' fuse against overload and short circuit and this fuse is also to give protection against indirect contact with a maximum disconnection time of 5 s.

If $I_b = 38$ A, $l = 90$ m, $t_a = 45°$C, $Z_E = 0.8$ ohm and the power factor of the load is 0.8 lagging, determine the minimum live and protective conductor cross-sectional area that can be used. The voltage drop is not to exceed $2\frac{1}{2}$% of the nominal voltage.

Answer

Because $I_n \geqslant I_b$ choose from the standard ratings of BS 88 Pt 2 fuses $I_n = 40$ A.

From Table 4B1, $C_g = 0.57$.

From Table 4C1, $C_a = 0.79$.

$$I_t = 40 \times \frac{1}{0.57} \times \frac{1}{0.79} \text{ A} = 88.8 \text{ A}$$

From Table 4D1A Column 4 it is found that the minimum conductor cross-sectional area that can be used is 25 mm^2 having $I_{ta} = 101$ A.

Check for voltage drop.

From Column 3 of Table 4D1B $(mV/A/m)_r = 1.8$ milliohms/m, $(mV/A/m)_x = 0.33$ milliohms/m, $(mV/A/m)_z = 1.8$ milliohms/m.

The voltage drop (using the simple approach) is given by:

$$\frac{1.8 \times 38 \times 90}{1000} \text{ V} = 6.2 \text{ V} = 2.7\% \text{ of } 230 \text{ V}$$

This is in excess of the permitted maximum but advantage should, in this case, be taken of the known power factor of the load. As this is 0.8, the value of $\sqrt{1 - \cos^2\phi} = 0.6$.

Thus the voltage drop is given by:

$$\frac{(0.8 \times 1.8) + (0.6 \times 0.33)}{1000} \times 38 \times 90 \text{ V} = 5.6 \text{ V} = 2.43\%$$

The circuit therefore *does* meet the limitation of voltage drop.

The next stage is to determine the minimum cross-sectional area for the circuit protective conductor.

From Table 41D the maximum value of Z_s that can be tolerated is 1.41 ohm.

Thus the maximum value of $(R_1 + R_2)$ that can be tolerated is $(1.41 - 0.8)$ ohm $= 0.61$ ohm.

The maximum value of $(R_1 + R_2)$/m that can be tolerated is therefore given by:

$$\frac{0.61 \times 1000}{90} \text{ milliohms/m} = 6.8 \text{ milliohms/m}$$

As Table 54C applies (because the circuit is grouped with other circuits) from Table 3.1 Column 4 the R_1/m for 25 mm^2 is 0.872 milliohms/m.

Thus the maximum R_2/m that can be tolerated is 5.93 milliohms/m and therefore the minimum cross-sectional area for the circuit protective conductor is $4\,mm^2$ having a value of 5.53 milliohms/m.

$$(R_1 + R_2) = \frac{(0.872 + 5.53) \times 90}{1000}\ ohm = 0.58\ ohm$$

$$Z_s = (0.8 + 0.58)\ ohm = 1.38\ ohm$$

The earth fault current I_{ef} is given by:

$$I_{ef} = \frac{230}{1.39}\ A = 167\ A$$

From the time/current characteristic in Appendix 3 of BS 7671 the disconnection time is found to be approximately 5 s. $k = 115$.

$$k^2S^2 = 115^2 \times 6^2\ A^2s = 476\,100\ A^2s$$

$$I_{ef}^2t = 167^2 \times 5\ A^2s = 139\,445\ A^2s$$

The circuit protective conductor is therefore thermally protected.

The test value for $(R_1 + R_2)$ at 20°C is found from Table 3.3 to be:

$$\frac{(0.727 + 4.61) \times 90}{1000}\ ohm = 0.480\ ohm$$

In this example there has been no point in considering whether the circuits are liable to simultaneous overload or not. The ratio I_b/I_n is $38/40 = 0.95$ and inspection of Figure 1.2 immediately shows that there is no real advantage to be gained as the reduction factor is very close to unity.

Example 6.3

A 230 V single-phase circuit is intended to be run in single-core non-armoured cables having 85°C rubber insulation and copper conductors, in conduit but not with the cables of other circuits. The circuit is to be protected against short circuit current only by a BS 88 'gG' fuse but this fuse is to provide protection against indirect contact, the maximum permitted disconnection time being 5 s.

If $I_b = 28\ A$, $t_a = 50°C$, $l = 24\ m$ and $Z_E = 0.35\ ohm$ what is the minimum cross-sectional area of the live and protective conductors that can be used? The short circuit current at the point of installation of the fuse has previously been calculated to be 4000 A and the voltage drop is not to exceed $2\frac{1}{2}\%$ of the nominal voltage.

Answer

As the fuse is not providing overload protection the design current I_b is used to determine the minimum cross-sectional area for the live conductors.

From Table 4C1, $C_a = 0.80$.

Thus:

$$I_t = 28 \times \frac{1}{0.80}\,A = 35\,A$$

From Table 4F1A Column 3 it is immediately seen that the minimum cross-sectional area for the live conductors appears to be 4 mm^2 having $I_{ta} = 40\,A$.

First check the voltage drop.

From Column 4 Table 4F1B the mV/A/m is 12 milliohms/m.

The voltage drop is given by:

$$\frac{12 \times 28 \times 24}{1000}\,V = 8.06\,V = 3.5\%$$

The circuit therefore does not meet the voltage drop limitation.

The maximum mV/A/m value which can be tolerated is given by

$$\frac{5750}{28 \times 24}\text{ milliohms/m} = 8.56\text{ milliohms/m}$$

Re-inspection of Column 4 shows that it is necessary to use conductors of 6 mm^2 cross-sectional area having mV/A/m of 7.7 milliohms/m and $I_{ta} = 52\,A$.

Now determine the minimum cross-sectional area of the circuit protective conductor. It is first necessary to determine I_n in order to establish the maximum earth fault loop impedance that can be tolerated. Whilst I_n must always be equal to or greater than I_b, Regulation 434–01–01 allows I_n to be *greater* than the effective current-carrying capacity of the conductor I_z when only short circuit protection is intended.

In the present case

$$I_z = 52 \times 0.8\,A = 41.6\,A$$

Let it be assumed that because of the characteristics of the load it is decided to use a fuse having $I_n = 50\,A$.

From Table 41D the maximum permitted Z_s is 1.09 ohm.

The maximum value of $(R_1 + R_2)$ is therefore:

$$(1.09 - 0.35)\text{ ohm} = 0.74\text{ ohm}$$

The maximum value of $(R_1 + R_2)/m$ that can be tolerated is given by

$$\frac{0.74 \times 1000}{24}\text{ milliohms/m} = 30.83\text{ milliohms/m}$$

Table 54B applies but if a check is to be made to see if the protective conductors can have a cross-sectional area of 1.5 mm^2, use Table 3.1, from which:

$$R_2/m = 15.2\text{ milliohms/m}$$

$$R_1/m = 3.88\text{ milliohms/m}$$

Thus $(R_1 + R_2)/m = 19.08$ milliohms/m and it would appear that $1.5\,mm^2$ cross-sectional area can be used for the circuit protective conductor. Thus:

$$(R_1 + R_2) = \frac{19.08 \times 24}{1000}\,\text{ohm} = 0.458\,\text{ohm}$$

$$Z_s = (0.35 + 0.458)\,\text{ohm} = 0.808\,\text{ohm}$$

The earth fault current is given by:

$$\frac{230}{0.808}\,A = 285\,A$$

From the time/current characteristic given in Appendix 3 of BS 7671 the disconnection time is 1.6 s.

From Table 54B, $k = 166$. Thus:

$$k^2S^2 = 166^2 \times 1.5^2\,A^2s = 62\,001\,A^2s$$

$$I_{ef}^2t = 285^2 \times 1.6\,A^2s = 129\,960\,A^2s$$

As $k^2S^2 < I_{ef}^2t$ the circuit does not meet the requirements of Regulation 543–01–03.

Try $2.5\,mm^2$ cross-sectional area for the protective conductor.

From Table 3.1 Column 5, $(R_1 + R_2)/m$ is found to be 11.6 milliohms/m.

Thus:

$$R_1 + R_2 = \frac{11.6 \times 24}{1000}\,\text{ohm} = 0.278\,\text{ohm}$$

$$Z_s = (0.35 + 0.278)\,\text{ohm} = 0.628\,\text{ohm}$$

$$I_{ef} = \frac{230}{0.628}\,A = 366\,A$$

From the time/current characteristic, $t = 0.6$ s. Thus:

$$k^2S^2 = 166^2 \times 2.5^2\,A^2s = 172\,225\,A^2s$$

$$I_{ef}^2t = 366^2 \times 0.6\,A^2s = 80\,374\,A^2s$$

As $k^2S^2 > I_{ef}^2t$ the circuit with a cpc of $2.5\,mm^2$ cross-sectional area meets the requirements of Regulation 543–01–03.

There remains the check that the circuit is adequately protected thermally under short-circuit conditions.

From Table 5.1, Column 4, $(R_1 + R_n)/m$ is found to be 7.76 milliohms/m so total phase/neutral impedance is given by:

$$\frac{230}{4000} + \left(\frac{7.76 \times 24}{1000}\right)\,\text{ohm} = 0.244\,\text{ohm}$$

The short circuit current is given by:

$$\frac{230}{0.244}\,\text{A} = 943\,\text{A}$$

From the time/current characteristics the disconnection time is found to be less than 0.01 s and it would be necessary to obtain the I^2t characteristic from the fuse manufacturer to check compliance with Regulation 434–03–03.

The test value for $(R_1 + R_2)$ is obtained from Table 3.3 and is given by:

$$\frac{(3.08 + 7.41) \times 24}{1000}\,\text{ohm} = 0.252\,\text{ohm}$$

In this example one step that should have been taken has been omitted. When it was found that $4\,\text{mm}^2$ live conductors apparently failed to meet the $2\frac{1}{2}\%$ voltage drop limitation a recalculation taking account of the actual conductor temperature should have been made.

With the $4\,\text{mm}^2$ cross-sectional area conductor:

$$\frac{I_b}{I_{ta}} = \frac{28}{45} = 0.62$$

From Figure 2.2 for this value of I_b/I_{ta} and $t_a = 50°C$ the reduction factor is found to be approximately 0.955.

The more correct voltage drop would therefore have been:

$$3.5 \times 0.955\% = 3.3\%$$

Thus, in this case, taking account of the actual conductor temperature has not avoided the need to increase the cross-sectional area to $6\,\text{mm}^2$.

Example 6.4

A 400 V three-phase four-wire circuit is to be taken from a sub-distribution board where the prospective short circuit current has been calculated to be 5500 A, the resistance component of the earth fault loop impedance is 0.2 ohm and the reactance component 0.08 ohm. The circuit is to be run in multicore non-armoured 70°C pvc-insulated (BS 6004) cable having copper conductors clipped direct and not grouped with the cables of other circuits. Protection against overload and short circuit is to be provided by BS 88 ‘gG’ fuses and these are also to give protection against indirect contact (maximum disconnection time 5 s).

If $t_a = 45°C$, $I_b = 41$ A and $l = 105$ m design the circuit for compliance with the relevant regulations, the maximum recommended voltage drop being $2\frac{1}{2}\%$ of the nominal voltage.

Answer

Select the nominal current of the fuse from the standard ratings.

$I_n = 50\,A.$

The only correction factor is that for ambient temperature.

From Table 4C1, $C_a = 0.79$. Thus:

$$I_t = 50 \times \frac{1}{0.79}\,A = 63.3\,A$$

From Table 4D2A, Column 7 the minimum cross-sectional area for the live conductors is found to be $16\,mm^2$ having $I_{ta} = 76\,A$. The cross-sectional area for the cpc is then $6\,mm^2$.

Now check the voltage drop.

From Table 4D2B, Column 4, mV/A/m = 2.4 milliohms/m.

$$\text{Voltage drop} = \frac{2.4 \times 105 \times 41}{1000}\,V = 10.3\,V$$

This is marginally greater than the permitted maximum voltage drop of 10 V.

From Figure 2.1 for $t_a = 45°C$ and $I_b/I_{ta} = 41/76 = 0.54$ it will be seen that the design mV/A/m can be taken to be approximately 0.96 times the tabulated value. Thus the 'corrected' voltage drop is given by

$$10.3 \times 0.96\,V = 9.9\,V$$

There is therefore no need to increase the conductor cross-sectional area.

A more accurate estimation of the voltage drop, if considered necessary, means first calculating t_1.

$$t_1 = 45 + \frac{41^2}{76^2}(70 - 30)°C = 56.6°C$$

so that:

$$\text{design mV/A/m} = \left(\frac{230 + 56.6}{230 + 70}\right) \times 2.4\,\text{milliohms/m} = 2.3\,\text{milliohms/m}$$

and

$$\text{voltage drop} = \frac{2.3 \times 105 \times 41}{1000}\,V = 9.9\,V$$

Now check that protection against indirect contact is obtained, using the simple approach.

From Table 3.1, Column 4 (because Table 54C applies):

$$(R_1 + R_2)/m = 5.08\,\text{milliohms/m}$$

$$R_1 + R_2 = \frac{5.08 \times 105}{1000}\,\text{ohm} = 0.53\,\text{ohm}$$

The earth fault loop impedance is given by:

$$Z_s = \sqrt{(0.2 + 0.53)^2 + 0.08^2} \text{ ohm} = 0.734 \text{ ohm}$$

From Table 41D it is found that the maximum value of Z_s for a disconnection time of 5 s is 1.09 ohm so the circuit meets this requirement.

It remains to check the thermal adequacy of the protective circuit.

Table 54C applies so that $k = 115$.

The earth fault current is given by:

$$\frac{230}{0.734} \text{ A} = 313 \text{ A}$$

From the time/current characteristic in Appendix 3 of BS 7671 the disconnection time is found to be approximately 0.9 s.

$$k^2S^2 = 115^2 \times 6^2 \text{ A}^2\text{s} = 476\,100 \text{ A}^2\text{s}$$

$$I_{ef}^2 t = 313^2 \times 0.9 \text{ A}^2\text{s} = 88\,172 \text{ A}^2\text{s}$$

The circuit conductors are adequately protected.

Because the fuses are intended to give both overload and short circuit protection there is no need to check that Regulation 434–01–03 has been met, Regulation 434–03–02 allows one to assume that it is met.

Example 6.5

A 400 V three-phase three-wire circuit is run in multicore armoured 70°C pvc-insulated cable having copper conductors. The cable is clipped direct and it is intended to use the armouring as the cpc. The circuit is to be protected against overload and short circuit by a T.P. Type B mcb. $I_b = 26$ A, $t_a = 40$°C and $l = 48$ m. The voltage drop must not exceed 4% of the nominal voltage and the disconnection time under earth fault conditions is not to exceed 5 s.

If $R_b = 0.06$ ohm, $X_b = 0.10$ ohm, $R_E = 0.12$ ohm and $X_E = 0.16$ ohm at the point of installation of the mcb, design the circuit for compliance with BS 7671.

Answer

Select the nominal current for the mcb from the standard ratings. Thus $I_n = 32$ A.

The symmetrical three-phase short circuit current at the point of installation of the mcb is given by:

$$\frac{230}{\sqrt{0.06^2 + 0.10^2}} \text{ A} = 1972 \text{ A}$$

A Type B mcb to BS EN 60989 having a 3000 A breaking capacity is therefore suitable.

First determine the cable conductor cross-sectional area.

From Table 4C1, $C_a = 0.87$.

$$I_t = 32 \times \frac{1}{0.87}\ \text{A} = 36.78\ \text{A}$$

From Table 4D4A Column 3 it is found that the minimum conductor cross-sectional area is 6 mm^2 having $I_{ta} = 42$ A.

Check for voltage drop.

From Table 4D4B Column 4, mV/A/m = 6.4 milliohms/m.

$$\text{Voltage drop} = \frac{26 \times 6.4 \times 48}{1000}\ \text{V} = 8\ \text{V} = 2\%$$

The voltage drop is therefore within the required maximum.

Now check the circuit under earth fault conditions.

From Table 3.11 the resistance of the steel wire armour is 4.6 milliohms/m at 20°C = 4.6 × 1.18 milliohms/m at 60°C = 5.43 milliohms/m. From Table 3.1 the resistance of a phase conductor is 3.70 milliohms/m.

Thus:

$$(R_1 + R_2) = \frac{(3.70 + 5.43)}{1000} \times 48\ \text{ohm} = 0.438\ \text{ohm}$$

$$Z_s = \sqrt{(0.438 + 0.12)^2 + 0.16^2}\ \text{ohm} = 0.58\ \text{ohm}$$

From Table 41B2 the maximum permitted value for Z_s is 1.5 ohm so the circuit complies.

It remains to check whether the cpc is adequate. In order to do this it is necessary to consult the cable manufacturer's data.

Assume that the armour cross-sectional area is found to be 37 mm^2.

From Table 54C, k for the conductor is 115, and from Table 54D, k for the armour is 51. To comply with Table 54G the armour cross-sectional area should be

$$\frac{115}{51} \times 6\ \text{mm}^2 \qquad \text{i.e. } 13.53\ \text{mm}^2.$$

The armour does comply and therefore there is no need to check that the circuit complies with the adiabatic equation.

Because the mcb is providing both overload and short circuit protection and has adequate breaking capacity there is no need to check compliance with the adiabatic equation under short circuit conditions.

One could, of course, provide many more examples of final circuits but it is felt that those which have been given show the approach that should be used.

When designing installations such as those shown in Figures 3.1 and 5.9 the first task facing the designer is to determine the most suitable locations for the sub-distribution boards. The route lengths of the distribution circuits can then be calculated and an assessment made of the design current in those circuits. That assessment requires a detailed knowledge of the complete installation including the operating conditions that will be encountered. Armed with this information the designer will be able to determine the appropriate diversity factors in order to obtain the design currents in the distribution circuits.

BS 7671 does not *demand* that diversity be taken into account but, for the economic design of an installation, diversity cannot be ignored. The diversity factors used must not be optimistic; an adequate margin of safety must always be allowed.

As indicated in Regulation 433–01–01 which gives the general requirement concerning overload protection, circuits shall be so designed that small overloads of long duration are unlikely to occur and this could well happen if the assessment of the design current in distribution circuits was based on optimistic diversity factors.

The designer should also give consideration to the possible future growth of the installation. The subject of diversity is, however, outside the scope of this present book.

The next stage in design is to apportion the permitted voltage drop between the distribution circuits and the final circuits. In this regard Regulation 525–01–02 which specifies a total voltage drop, i.e. between the origin of the installation and the current-using equipment, of 4% of the nominal voltage, has a 'deemed to satisfy' status.

Regulation 525–01–01 which is the basic regulation in effect allows more than 4% provided that the resulting voltage at the terminals of the current-using equipment is not less than any lower limit specified in the relevant British Standard or, where there is no British Standard, is such as not to impair the safe functioning of that equipment. Having decided on the voltage drop which can be permitted in the final circuits the detailed design of those circuits can be undertaken. At this stage these designs are provisional because it may be found that the calculated conductor cross-sectional areas are unduly large. For example, it might be found that the conductors could not be accommodated in the equipment or accessory terminations.

It may therefore be necessary to examine the number of final circuits grouped together, whether circuits running through areas of high ambient temperature or in contact with thermally insulating material could be re-routed and so on. These changes should be deferred until after the distribution circuits have been designed because if their voltage drops are well within their allocated values some increase in the voltage drops in some of the final circuits may be possible.

Here let it be assumed that the final circuits have been designed and it remains to calculate the conductor cross-sectional areas for the distribution circuits, to check the voltage drops and to check compliance with the requirements for protection against indirect contact both as regards meeting maximum disconnection times and the thermal adequacy of the circuit live and protective conductors under earth fault conditions.

Furthermore the prospective short circuit currents at the distribution boards have to be calculated in order to check that there is adequate co-ordination between protective devices and, unless Regulation 434–03–02 is invoked, the adequacy of the thermal protection of the circuit conductors.

The design approach consists of calculating the appropriate circuit resistances and impedances but, as already indicated in earlier chapters, these are based on different temperatures depending on which aspect is being checked.

Example 6.6

The installation as shown in Figure 6.1 is fed from a 200 kVA three-phase delta/star transformer having a nominal secondary voltage (on load) of 400 V between phases and 230 V phase-to-neutral, the neutral being distributed throughout the installation. The resistance per phase of the transformer is 0.016 ohm and the reactance per phase is 0.04 ohm. The ambient temperature throughout the installation can be assumed to be 30°C.

The estimated current in the distribution circuits are as shown in the figure together with the nominal ratings of the fused switches (fuses to BS 88 'gG') intended in all cases to give both overload and short circuit protection to the distribution circuits concerned. The distribution circuits are to be run in multicore armoured 70°C pvc-insulated cables, clipped direct, and the estimated route lengths are as shown. Each circuit is not grouped with other circuits. The voltage drop from the transformer to distribution board A is not to exceed 1.5 V line-to-line and for each of the distribution circuits from distribution board A to the sub-distribution boards B and C the voltage drop is not to exceed 3 V line-to-line.

Determine the conductor cross-sectional areas of the distribution cables and check compliance with the requirements for overcurrent protection and protection against indirect contact.

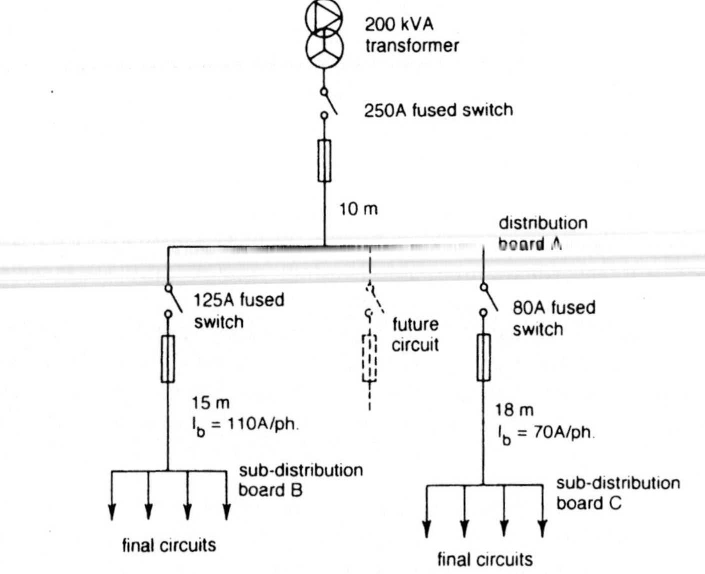

Figure 6.1 Installation schematic diagram for Example 6.6.

Answer

First determine the conductor cross-sectional areas. As the ambient temperature can be taken to be 30°C and there is no grouping there are no correction factors to be applied.

Thus from the transformer to distribution board A:

$$I_t = I_n = 250\,\text{A}$$

From Table 4D4A, Column 3 it is found that the minimum conductor cross-sectional area that can be used is 120 mm² having $I_{ta} = 267$ A.

From distribution board A to sub-distribution board B, again from Table 4D4A Column 3, the minimum conductor cross-sectional area is 35 mm² having $I_{ta} = 125$ A, i.e. exactly the same as I_n.

From distribution board A to sub-distribution board C, $I_t = I_n = 80$ A and from Table 4D4A Column 3 the minimum conductor cross-sectional area for this cable is 25 mm² having $I_{ta} = 102$ A.

At this point in design the above conductor cross-sectional areas give compliance only with the requirements for overload protection. As the prospective short circuit circuits have not yet been determined it is too early to claim that the overcurrent protective devices have adequate breaking capacity and it is not possible to invoke Regulation 434–03–02, i.e. it is not possible to claim the circuits concerned are adequately protected against short circuit.

But first check the voltage drops.

From Column 4 of Table 4D4B the following values of mV/A/m are obtained (all in milliohms/m):

	$(\text{mV/A/m})_r$	$(\text{mV/A/m})_x$	$(\text{mV/A/m})_z$
25 mm²	1.50	0.145	1.50
35 mm²	1.10	0.145	1.10
120 mm²	0.33	0.135	0.35

As no indication is given of the power factor in any of the distribution circuits the $(\text{mV/A/m})_z$ values are used in each case and the voltage drops are calculated as follows.

From the transformer to distribution board A:

$$\text{voltage drop} = \frac{0.35 \times 10 \times 250}{1000}\,\text{V} = 0.88\,\text{V}$$

(Here $I_n = 250$ A has been used because, as indicated, there is a further distribution circuit to be taken from distribution board A.)

From distribution board A to sub-distribution board B:

$$\text{voltage drop} = \frac{1.10 \times 15 \times 110}{1000}\,\text{V} = 1.82\,\text{V}$$

From distribution board A to sub-distribution board C:

$$\text{voltage drop} = \frac{1.50 \times 18 \times 70}{1000}\,\text{V} = 1.89\,\text{V}$$

All the circuits therefore meet the specified limitation in voltage drop.

Now determine the prospective short circuit currents.

At the terminals of the transformer:

$$I_{pscc} = \frac{230}{\sqrt{0.016^2 + 0.04^2}}\,\text{A} = 5339\,\text{A}$$

As a matter of interest another way of determining this prospective short circuit current is to use the %Z of the transformer, this being given on the transformer rating plate and:

$$I_{pscc} = \frac{\text{kVA rating} \times 1000}{\sqrt{3} \times U_L} \times \frac{100}{\%Z}\,\text{A}$$

The other important relationships with transformers are:

$$\text{Z phase} = \frac{U_L^2 \times \%Z}{\text{kVA rating} \times 100\,000}\,\text{ohm}$$

$$\text{R phase} = \frac{U_L^2 \times \%R}{\text{kVA rating} \times 100\,000}\,\text{ohm}$$

$$\text{X phase} = \frac{U_L^2 \times \%X}{\text{kVA rating} \times 100\,000}\,\text{ohm}$$

with %R being the percentage resistance
%X being the percentage reactance
U_L being the line-to-line voltage.

In the present case:

$$Z_{ph} = \sqrt{0.016^2 + 0.04^2}\,\text{ohm} = 0.0431\,\text{ohm}$$

$$\%Z = \frac{0.0431 \times 200 \times 100\,000}{400^2} = 5.4\%$$

Using this to obtain the prospective short-circuit current:

$$I_{pscc} = \frac{200 \times 1000}{\sqrt{3} \times 400} \times \frac{100}{5.4}\,\text{A} = 5346\,\text{A}$$

Because the prospective short circuit is only 5346 A and the protective devices for the two circuits from distribution board A to the sub-distribution boards B and C are BS 88 'gG' fuses having breaking capacities considerably in excess of 5346 A there is strictly no need to determine the prospective short circuit current at distribution board A.

In the present case, because all the fuses are providing overload protection as well as short circuit protection and have adequate breaking capacity, Regulation 434–03–02 applies so that it is not necessary to check the thermal capabilities of the cables under short circuit conditions using the adiabatic equation of Regulation 434–03–03.

Notwithstanding the above remarks and simply to illustrate the method to be used, the prospective short circuit currents at the three distribution boards are now calculated. In order to calculate these prospective short circuit currents it is necessary to know the resistance and reactance of one phase conductor of the circuit concerned at 20°C (or some other temperature, if appropriate, because the premise is that the fault could occur in a 'cold' installation). The resistance and reactance of the neutral conductor is not needed.

The necessary resistance/metre and reactance/metre values may be obtained directly from the cable manufacturer's data because these sometimes give these values at 20°C.

Alternatively the mV/A/m values given in Appendix 4 of BS 7671 can be used 'correcting' them to 20°C.

Using the latter method, for the main incoming circuit:

$$R_1 = \frac{(\text{mV/A/m})_r \times 1 \times (230 + 20)}{\sqrt{3} \times 1000 \times (230 + 70)} \text{ ohm}$$

$$= \frac{0.33 \times 10 \times 250}{\sqrt{3} \times 1000 \times 300} \text{ ohm}$$

$$= 0.0016 \text{ ohm}$$

$$X_1 = \frac{(\text{mV/A/m})_x \times 1}{\sqrt{3} \times 1000} \text{ ohm}$$

$$= \frac{0.135 \times 10}{\sqrt{3} \times 1000} \text{ ohm}$$

$$= 0.0008 \text{ ohm}$$

The prospective short circuit current at distribution board A is given by:

$$I_{pscc} = \frac{230}{\sqrt{(0.016 + 0.0016)^2 + (0.04 + 0.0008)^2}} \text{ A} = 5176 \text{ A}$$

Similarly for the circuit from distribution board A to sub-distribution board B:

$$R_1 = \frac{1.10 \times 15 \times 250}{\sqrt{3} \times 1000 \times 300} \text{ ohm}$$

$$= 0.0079 \text{ ohm}$$

$$X_1 = \frac{0.145 \times 15}{\sqrt{3} \times 1000} \text{ ohm}$$

$$= 0.0013 \text{ ohm}$$

The prospective short circuit current at distribution board B is given by:

$$I_{pscc} = \frac{230}{\sqrt{(0.016 + 0.0016 + 0.0079)^2 + (0.04 + 0.0008 + 0.0013)^2}}\ A$$

$$= 4673\ A$$

For the circuit from distribution board A to sub-distribution board C:

$$R_l = \frac{1.5 \times 18 \times 250}{\sqrt{3} \times 1000 \times 300}\ ohm$$

$$= 0.013\ ohm$$

$$X_l = \frac{0.145 \times 18}{\sqrt{3} \times 1000}\ ohm$$

$$= 0.0015\ ohm$$

The prospective short circuit current at distribution board C is given by:

$$I_{pscc} = \frac{230}{\sqrt{(0.016 + 0.0016 + 0.013)^2 + (0.04 + 0.0008 + 0.0015)^2}}\ A$$

$$= 4405\ A$$

The next stage in design is to check the earth fault loop impedances for the circuits concerned and, at the same time, whether the circuit conductors are thermally protected. To calculate the earth fault loop impedance at distribution board A it is necessary to use the value of R_l for the main distribution circuit at the assumed temperature under earth fault conditions, namely 70°C.

Again the $(mV/A/m)_r$ value can be used:

$$R_l = \frac{0.33 \times 10}{\sqrt{3} \times 1000}\ ohm = 0.0019\ ohm$$

$$X_l = 0.0008\ ohm\ \text{(as before)}$$

The resistance and reactance per metre for the steel wire armour are obtained from Table 3.11, Columns 6 and 7 respectively.

The resistance value has to be corrected to 60°C by multiplying by 1.18. Thus:

$$R_2 = \frac{0.71 \times 10 \times 1.18}{1000}\ ohm = 0.0084\ ohm$$

$$X_2 = \frac{0.3 \times 10}{1000} = 0.003\ ohm$$

The earth fault loop impedance at distribution board A is given by:

$$Z_s = \sqrt{(0.016 + 0.0019 + 0.0084)^2 + (0.04 + 0.0008 + 0.003)^2}\ \text{ohm}$$

$$= \sqrt{(0.0263)^2 + (0.0438)^2}\ \text{ohm}$$

$$= 0.051\ \text{ohm}$$

The time/current characteristics in Appendix 3 of BS 7671 do not include that for 250 A fuses, hence it is necessary to obtain it from the fuse manufacturer.

A typical earth fault current to give 5 s disconnection time is 1500 A giving a maximum earth fault loop impedance of 0.15 ohm. The main incoming circuit complies as regards disconnection time.

From the cable manufacturer's data it is found that the gross cross-sectional area of the armour wires is 221 mm^2.

First check if this cross-sectional area complies with Table 54G of BS 7671.

According to that table the cross-sectional area should be at least:

$$\frac{115}{51} \times \frac{120}{2}\ \text{mm}^2 = 135.3\ \text{mm}^2$$

The cable therefore complies with Table 54G and there is no need to check further the thermal capability of the conductors under earth fault conditions.

To determine the earth fault loop impedance at distribution boards B and C it is first necessary to recalculate the resistance of the phase conductor of the main distribution circuit at the normal operating temperature (70°C for pvc). To do this the $(\text{mV/A/m})_r$ value is used without a correction factor. The reactance remains the same.

The armouring temperature was assumed not to rise significantly under earth fault conditions and was previously calculated at the normal operating temperature of 60°C (i.e. 10°C less than the conductor operating temperature). Thus no change is needed.

$$R_1 = \frac{0.33 \times 10}{\sqrt{3} \times 1000}\ \text{ohm} = 0.0019\ \text{ohm}$$

$$X_1 = 0.0008\ \text{ohm (as before)}$$

$$R_2 = 0.0084\ \text{ohm (as before)}$$

$$X_2 = 0.003\ \text{ohm (as before)}$$

For the distribution circuit from distribution board A to sub-distribution board B:

$$R_1 = \frac{1.10 \times 15}{\sqrt{3} \times 1000} \text{ ohm}$$

$$= 0.0095 \text{ ohm}$$

X_1 as before $= 0.0013$ ohm

From Table 3.11, Columns 6 and 7:

$$R_2 = \frac{1.9 \times 15 \times 1.18}{1000} \text{ ohm}$$

$$= 0.0336 \text{ ohm}$$

$X_2 = 0$ (i.e. negligible)

The earth fault loop impedance at sub-distribution board B is then obtained as follows.

Its resistance component is made up of:

Internal resistance of the transformer 0.016 ohm
R_1 for phase conductor of main distribution circuit (at 70°C) 0.0019 ohm
R_2 for armouring of main distribution circuit (at 60°C) 0.0084 ohm
R_1 for phase conductor of distribution circuit to board B (at 70°C) 0.0095 ohm
R_2 for armouring of distribution cable to board B (at 60°C) 0.0336 ohm

Total resistance 0.0694 ohm

Its reactance component is made up of:

Internal reactance of the transformer 0.04 ohm
X_1 for phase conductor of main distribution circuit 0.0008 ohm
X_2 for armouring of main distribution circuit 0.003 ohm
X_1 for phase conductor of distribution circuit to board B 0.0013 ohm
X_2 for armouring of distribution circuit to board B is negligible 0

Total reactance 0.0451 ohm

Thus:

$$Z_s = \sqrt{0.0694^2 + 0.0451^2} \text{ ohm} = 0.083 \text{ ohm}$$

From Table 41D of BS 7671 the maximum earth fault loop impedance for 5 s disconnection is 0.35 ohm so the circuit complies.

In order to be able to apply Table 54G the cross-sectional area for the armour wires has to be at least

$$\frac{115}{51} \times 16 \text{ mm}^2 = 36 \text{ mm}^2$$

A check of the cable manufacturer's data has to be made and it will be found that Table 54G will be met, a typical value being 85 mm^2 for the armour cross-sectional area.

For the distribution circuit from distribution board A to sub-distribution board C:

$$R_1 = \frac{1.5 \times 18}{\sqrt{3} \times 1000} \text{ ohm}$$

$$= 0.016 \text{ ohm}$$

$$X_1 \text{ as before} = 0.0015 \text{ ohm}$$

From Table 3.11, Columns 6 and 7:

$$R_2 = \frac{2.1 \times 18 \times 1.18}{1000} \text{ ohm}$$

$$= 0.0446 \text{ ohm}$$

$$X_2 = 0 \text{ (i.e. negligible)}$$

The earth fault loop impedance at sub-distribution board C is then obtained as follows. Its resistance component is made up of:

Internal resistance of the transformer	0.016 ohm
R_1 for phase conductor of main distribution circuit (at 70°C)	0.0019 ohm
R_2 for armouring of main distribution circuit (at 60°C)	0.0084 ohm
R_1 for phase conductor of distribution circuit to board C (at 70°C)	0.016 ohm
R_2 for armouring of distribution cable to board C (at 60°C)	0.0446 ohm
Total resistance	0.0869 ohm

Its reactance component is made up of:

Internal reactance of the transformer	0.04 ohm
X_1 for phase conductor of main distribution circuit	0.0008 ohm
X_2 for armouring of main distribution circuit	0.003 ohm
X_1 for phase conductor of distribution circuit to board C	0.0015 ohm
X_2 for armouring of distribution cable to board C is negligible	0
Total reactance	0.0453 ohm

Thus:

$$Z_s = \sqrt{0.0869^2 + 0.0453^2} = 0.098 \text{ ohm}$$

This is well within the maximum value of Z_s for an 80 A BS 88 'gG' fuse given in Table 41D, namely 0.6 ohm.

Again it will be found that Table 54G is met and there is no need to check further the thermal capacity of the armouring under earth fault conditions.

Appendix

The Touch Voltage Concept

See also: *Touch Voltages in Electrical Installations*, B. D. Jenkins (Blackwell Science, 1993).

The object of BS 7671 includes the provision of requirements concerning protection against electric shock.

Two forms of electric shock are recognised: the first of these is termed 'direct contact' which is contact of persons or livestock with live parts and the second is 'indirect contact', which is contact of persons or livestock with exposed-conductive-parts made live by a fault.

The touch voltage concept is concerned with the latter form of electric shock. It is used to determine the magnitude of the voltage to which the person at risk would be subjected in the event of an earth fault occurring in an installation. By assuming values of body resistance the touch voltage concept can be extended to give an indication of the severity of the electric shock that could be experienced by that person.

A study of the touch voltage concept, which is a very simple concept, gives the installation designer, and others concerned with electrical installations, a fuller understanding of the requirements given in BS 7671 for protection against indirect contact.

One often hears or reads the statement that provided an item of electrical equipment (intended to be earthed) is properly earthed it is not possible for a person to receive an electric shock in the event of a fault. The touch voltage concept shows such a statement to be totally incorrect.

The connection of all exposed- and extraneous-conductive-parts, either directly or indirectly, to a common terminal, namely the main earthing terminal of the installation, leads to the creation of touch voltages in the event of an earth fault and hence to the shock risk. In a correctly designed and erected electrical installation the shock risk from indirect contact is *not* eliminated. Where the protective measure is automatic disconnection of supply then in the event of an earth fault, the speed of disconnection should be such that should the person at risk experience an electric shock it will not be a harmful one.

Figure A.1 is the basic simplified schematic diagram for a TN–S system comprising a source of energy and an installation.

E_o = induced emf (to earth of the source), V
Z_i = internal impedance of the source, ohm
Z_1 = impedance of the circuit phase conductor, ohm
Z_2 = impedance of the circuit protective conductor, ohm
Z_3 = impedance of the supply cable protective conductor, ohm
Z_4 = impedance of the supply cable phase conductor, ohm

As shown in the figure an earth fault has occurred in equipment A, the phase conductor coming into contact with the metallic enclosure of the equipment.

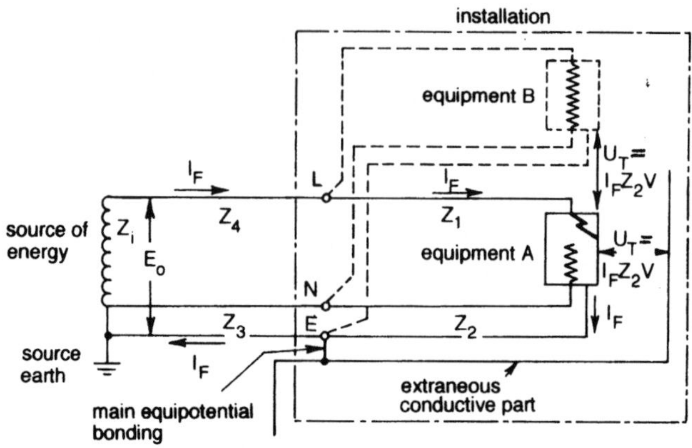

Figure A.1 Basic schematic diagram for a TN–S system showing touch voltages created by an earth fault.

The assumption is made that the fault itself is of negligible impedance and is accompanied by an open circuit in the equipment so that no part of the equipment impedance is in the earth fault loop.

The earth fault current, I_F, is given by:

$$I_F = \frac{E_o}{Z_i + Z_1 + Z_2 + Z_3 + Z_4} A = \frac{E_o}{Z_s} A \quad (1)$$

Z_s is the earth fault loop impedance.

Now $Z_i + Z_3 + Z_4 = Z_E$ i.e. that part of Z_s which is external to the installation, so that:

$$I_F = \frac{E_o}{Z_E + Z_1 + Z_2} A \quad (2)$$

When the circuit conductors have a cross-sectional area of less than 35 mm^2, their resistance may be used instead of their impedance and a sufficient degree of practical accuracy is obtained if U_o, the nominal voltage to earth of the supply, is used instead of E_o. Thus:

$$I_F = \frac{U_o}{Z_E + R_1 + R_2} A \quad (3)$$

The introduction to Appendix 3 of BS 7671 indicates that it may be more appropriate to use U_{oc}, the open circuit voltage at the distribution transformer, in place of U_o in Eqn. (3). U_o has been used in this Appendix to be consistent with the main body of this book. If the open circuit voltage is known, then the following equations can be modified by substituting U_{oc} for U_o.

If consideration is limited to the case where the person at risk is inside the shock protection zone (i.e. the so-called equipotential zone created by the main equipotential bonding) and therefore is not directly in contact with the general mass of earth, in the event of an earth fault as shown in Figure A.1, if that person simultaneously touches the extraneous-conductive-part and the exposed-conductive-part

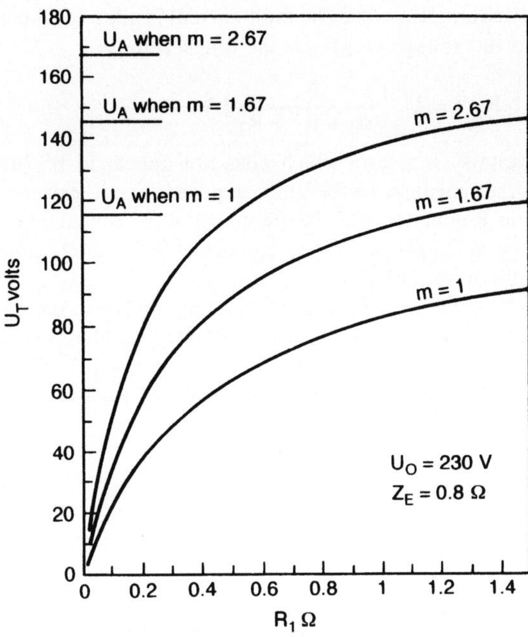

Figure A.2 Touch voltage against resistance of phase conductor (R_1) for various m values when $U_o = 230\,V$ and $Z_E = 0.80$ ohm.

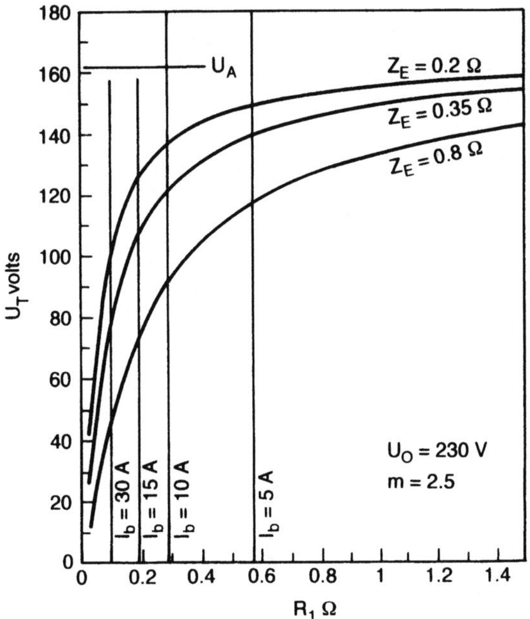

Figure A.3 Touch voltage against resistance of phase conductor (R_1) for various Z_E values when $U_o = 230\,V$ and $m = 2.5$. The vertical lines for various values of I_b correspond to a permitted voltage drop of $2\frac{1}{2}\%$.

of the faulty equipment while the fault current is allowed to persist, he or she will be subjected to the voltage U_T. U_T is the touch voltage:

$$U_T = I_F R_2\,V = U_o\left(\frac{R_2}{Z_E + R_1 + R_2}\right) V \qquad (4)$$

'Touch voltage' is a term which does not appear in BS 7671 but is 'the voltage between simultaneously accessible exposed- and extraneous-conductive-parts' referred to in Regulation 413–02–04 which is the fundamental requirement for the protective measure against indirect contact using earthed equipotential bonding and automatic disconnection of supply.

Again for conductors of cross-sectional area not exceeding $35\,mm^2$, if the phase and protective conductors of a circuit are of the same material and are run over the same route:

$$\frac{R_2}{R_1} = \frac{A_1}{A_2} = m$$

where A_1 = cross-sectional area of the phase conductor, in mm^2
A_2 = cross-sectional area of the protective conductor, in mm^2

Equation (4) then becomes:

$$U_T = U_o \frac{m}{(Z_E/R_1) + 1 + m}\,V \qquad (5)$$

Equation (5) is, therefore, the basic touch voltage equation.

From it a number of important points emerge.

(1) For a particular value of Z_E and of 'm', as R_1 increases, the denominator decreases so that U_T increases. In other words, for a particular circuit, the maximum touch voltage occurs when the earth fault is at the remote end (for a ring circuit, when the earth fault is at the mid-point).
(2) For a particular value of R_1 and of 'm', as Z_E decreases, the denominator decreases so that U_T increases. In other words, for a particular circuit the touch voltage can approach but never quite reach the value obtained by putting $Z_E = 0$ in equation (5) (the basic touch voltage equation).

This gives the asymptotic value denoted here by U_A. Thus:

$$U_A = U_o\left(\frac{m}{m+1}\right) V$$

Table A.1 Values of m and U_A/U_o for flat twin cable.

Phase conductor cross-sectional area mm^2	Protective conductor cross-sectional area mm^2	m	$\frac{U_A}{U_o}$	U_A when $U_o = 230\,V$ volts
1	1	1	0.5	115
1.5	1	1.5	0.6	138
2.5	1.5	1.67	0.625	144
4	1.5	2.67	0.728	167
6	2.5	2.4	0.706	162
10	4	2.5	0.714	164
16	6	2.67	0.728	167

Table A.1 is based on the flat twin-core and three-core with cpc 70°C pvc-insulated and sheathed cables to BS 6004, but it is equally applicable to single-core cables.

Figure A.2 shows a family of curves, each curve for a particular value of m, of U_T plotted against R_1 when $U_o = 230$ V and $Z_E = 0.8$ ohm. Figure A.3 shows a family of curves, each curve for a particular value of Z_E when $U_o = 230$ V and m = 2.5 and superimposed on this family are vertical lines giving values of R_1 for different values of I_b to give a $2\frac{1}{2}$% voltage drop in a 230 V single-phase circuit.

Returning for the moment to Figure A.1 it will be seen that the touch voltage, $U_T = I_F Z_2$ V also exists between the exposed conductive parts of the faulty equipment and those of the healthy equipment fed by another circuit because of the common connection of the protective conductors of both circuits to the main earthing terminal (E) of the installation.

Figure A.4 shows the schematic diagram for a multi-outlet radial circuit and the earth fault is at the remote outlet. x_1, x_2 and x_3 are the fractional distances of the protective conductor ($x_1 + x_2 + x_3 = 1$). It will be seen that touch voltages of different magnitudes exist, even between exposed conductive parts of two healthy equipments, but the maximum value occurs between the exposed-conductive-part of the faulty equipment and extraneous-conductive-parts.

Thus, in the event of an earth fault, the zone created by the main equipotential bonding is far from 'equipotential' hence the preference for calling the zone the 'protected' zone. The zone is truly equipotential only when the earth fault occurs outside the zone. When this happens the main earthing terminal will take up some potential with respect to true earth and all the exposed- and extraneous-conductive-parts will take up that potential.

Having considered the magnitude of the touch voltages occurring in an installation, at least those related to a circuit connected directly at the origin of the installation and not via, for example, a sub-distribution board, there remains the aspect of what is considered to be the time for which those touch voltages can persist without causing danger.

Based on the data given in the IEC Report – Publication 479 *Effects of current passing through the human body* – (identical to the BSI Publication PD6519) and using certain values for the resistance of the human body, the current/time zones of that publication were translated into the two touch voltage duration curves as shown in Figure A.5 for 50 Hz a.c.

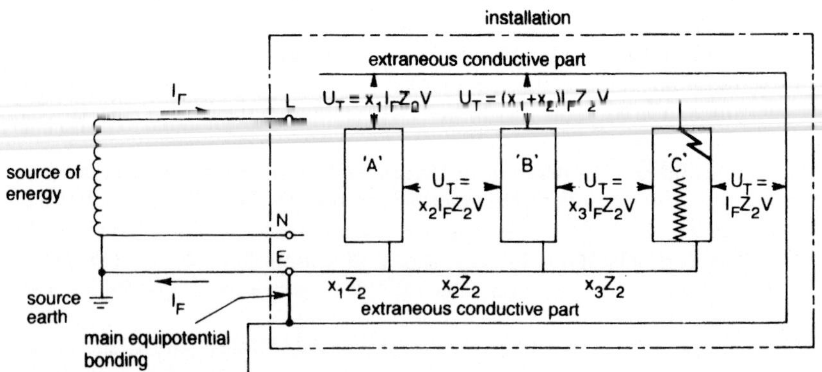

NOTE: Only the circuit protective conductor is shown

Figure A.4 Basic schematic diagram for a multi-outlet circuit in a TN–S system showing touch voltages created by an earth fault.

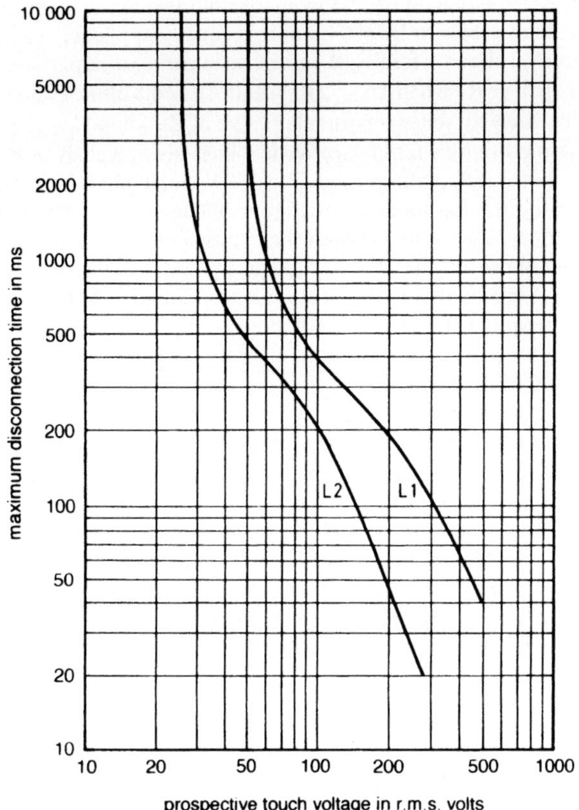

Figure A.5 IEC touch voltage duration curves.

The curve L1 relates to Condition 1, defined as normally dry situations, where the surface on which the person at risk is standing presents some resistance (to the general mass of earth) and that person is assumed to have dry or moist skin.

The curve L2 relates to Condition 2, defined as wet locations, where that surface does not present any resistance and the person is assumed to have wet skin.

The international committee IEC TC 64 decided not to adopt these touch voltage duration curves into the international Chapter 41 but adopted the maximum disconnection time used in the 15th Edition of the Wiring Regulations of 0.4 s for circuits having $U_o = 240$ V.

Limiting consideration to Condition 1 it will be noted that when the touch voltage is 50 V the disconnection time can be 5 s *or greater*. In other words, if the touch voltage is 50 V or less automatic disconnection of the supply is not required from consideration of electric shock. Disconnection is required, however, from thermal considerations. This value of 50 V is known as the conventional touch voltage limit (U_L).

The earth fault loop impedance, Z_s, at the remote end of a radial circuit (or at the mid-point of a ring circuit) determines the magnitude of the earth fault current, I_F, which in turn determines the time of disconnection of the overcurrent protective device being used to provide protection against indirect contact. The maximum touch voltage which can be tolerated for that disconnection time is then obtained

from the relevant touch voltage duration curve and it is an easy matter to check whether the resistance of the circuit protective conductor (R_2) is such that the calculated touch voltage is less.

However, there is a very simple graphical method that can be used, developed by one of the authors of this book some years ago, which requires the production of what are called 'impedance characteristics' for fuses and miniature circuit breakers. These are obtained in the following manner.

Figure A.6 shows the time/current characteristic for an HBC fuse and the touch voltage duration curve but solely for the purpose of explanation the mirror image of the latter has been used.

Take any value of earth fault current such as I_{F1}. This corresponds to an earth fault loop impedance Z_{S1} given by:

$$Z_{S1} = \frac{U_o}{I_{F1}} \text{ ohm}$$

From the time/current characteristic obtain the corresponding disconnection time t_1 and then from the touch voltage duration curve obtain the maximum value of the touch voltage U_{T1} which can be allowed to persist for this time t_1.

The maximum permitted value of the circuit protective conductor resistance is then given by:

$$R_{21} = \frac{U_{T1}}{I_{F_1}}$$

On plain graph paper plot R_{21} against Z_S.

This procedure is then repeated for other chosen values of earth fault current.

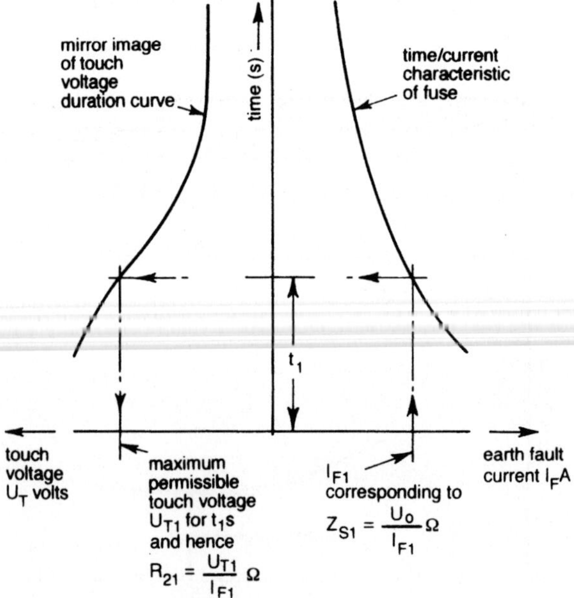

Figure A.6 Derivation of an impedance characteristic.

In producing an impedance characteristic there is no point in considering values of the disconnection time t below that value on the touch voltage duration curve corresponding to U_T being equal to U_o.

For values of Z_s greater than that corresponding to a 5 s disconnection time the maximum permissible value of R_2 is no longer dependent on the time/current characteristic of the protective device concerned but varies linearly with Z_s and is given by:

$$R_2 = \frac{U_L}{U_o} \times Z_s \text{ ohm}$$

For Condition 1, $U_L = 50$ V and if $U_o = 230$ V, $R_2 = 0.217\,Z_s$.

Using the above procedure the following impedance characteristics have been developed, all for $U_o = 230$ V and Condition 1.

Figure A.7 25 A BS 88 Fuse
Figure A.8 30 A BS 3036 Fuse
Figure A.9 20 A BS EN 60898 Type B mcb

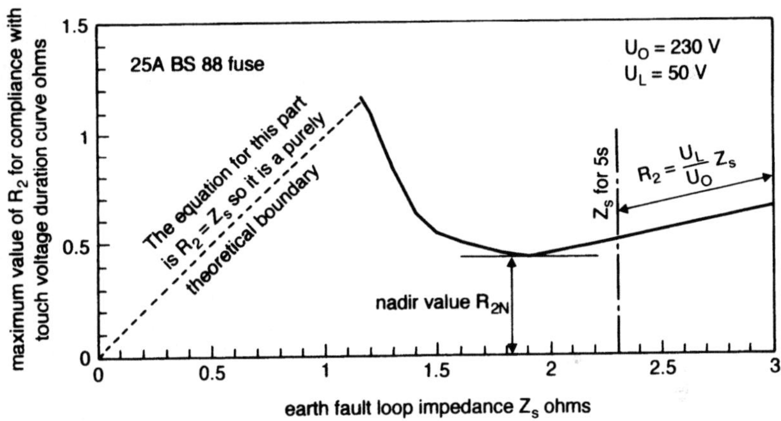

Figure A.7 Impedance characteristic for a BS 88 'gG' fuse.

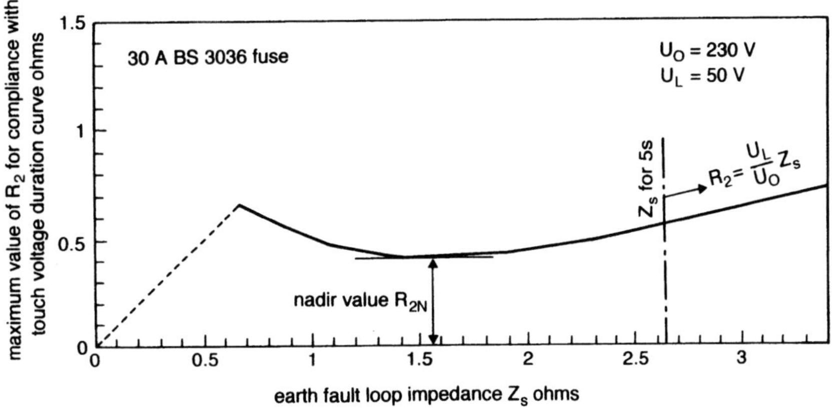

Figure A.8 Impedance characteristic for a BS 3036 semi-enclosed fuse.

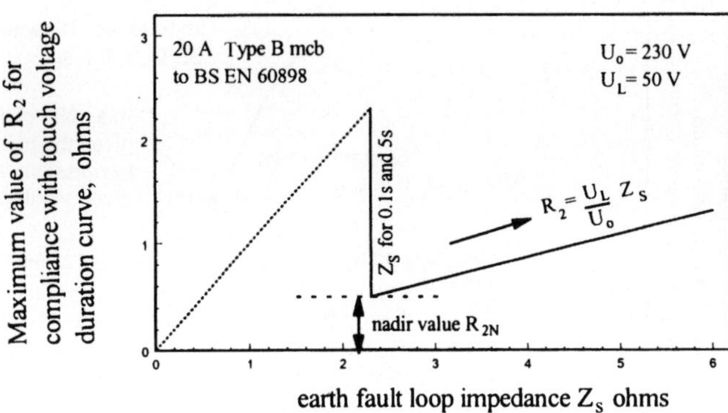

Figure A.9 Impedance characteristics for a Type B mcb.

When the resistance of the cpc is limited to this value the 0.4 s maximum disconnection time for socket outlet circuits can be increased to 5 s. But it also means that even if the earth fault itself has some impedance or if part of the load impedance is on the earth fault path the circuit will still comply with the touch voltage curve.

The nadir value for mcbs is shown in Figure A.9 as this is included in Table 41C of BS 7671.

Examination of the time/current characteristics for mcbs given in Appendix 3 of BS 7671 shows that for a particular type and rating of mcb the prospective currents for 0.1 s, 0.4 s, and 5 s disconnection times are of one value. Thus there is no practical use of the nadir value for mcbs because the maximum permitted earth fault loop impedances for these times are of one value. In any event the 0.1 s disconnection time is below the 0.16 s maximum allowed in the touch voltage duration curve for a touch voltage of 230 V and within the so-called equipotential zone of an installation having $U_o = 230$ V the touch voltage cannot, in fact, attain that value.

Figures A.10 and A.11 show the two basic ways in which the impedance characteristic may be used for design purposes. Figure A.10 is used for cases where the 'm'

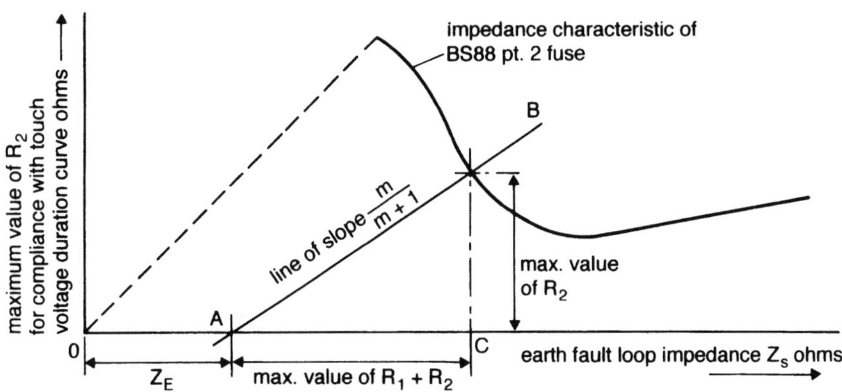

Figure A.10 Using impedance characteristic to determine maximum tolerable value of $(R_1 + R_2)$ and R_2.

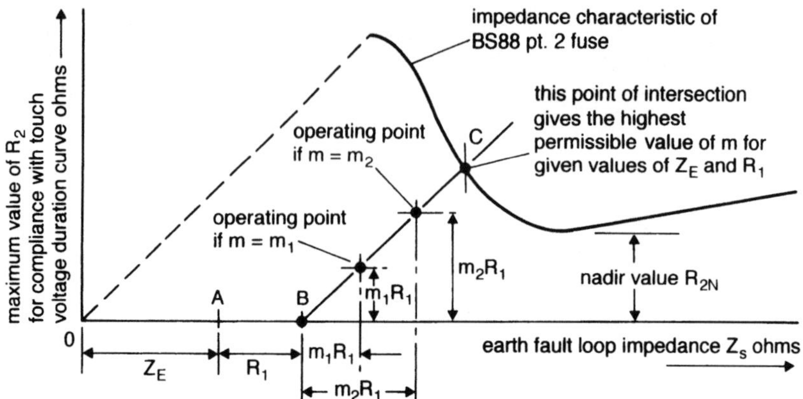

Figure A.11 Using impedance characteristic to determine maximum tolerable value of m.

value is known, e.g. in the flat two-core and three-core pvc-insulated and sheathed cables to BS 6004. The line AB is the locus of operation and where it meets the impedance characteristic this gives the maximum value of $(R_1 + R_2)$ that can be tolerated and this can then be translated into maximum circuit length. Figure A.11 is used for cases where Z_E and R_1 are known and one wishes to determine the maximum value of 'm' which can be tolerated, i.e. the minimum cross-sectional area for the circuit protective conductor. The line BC is the locus of operation and its point of intersection with the impedance characteristic gives the maximum tolerable value of 'm'.

It is necessary in both cases to check that the circuit meets the thermal requirements of Chapter 54 of BS 7671.

The particular advantage of using impedance characteristics is that there is no need to calculate the touch voltage as such, but various values of touch voltage can be constructed on the impedance characteristic as indicated in Figure A.12.

For a touch voltage of U_T the slope of the corresponding line is given by U_T/U_o. It would therefore appear that the impedance characteristic (and the touch voltage

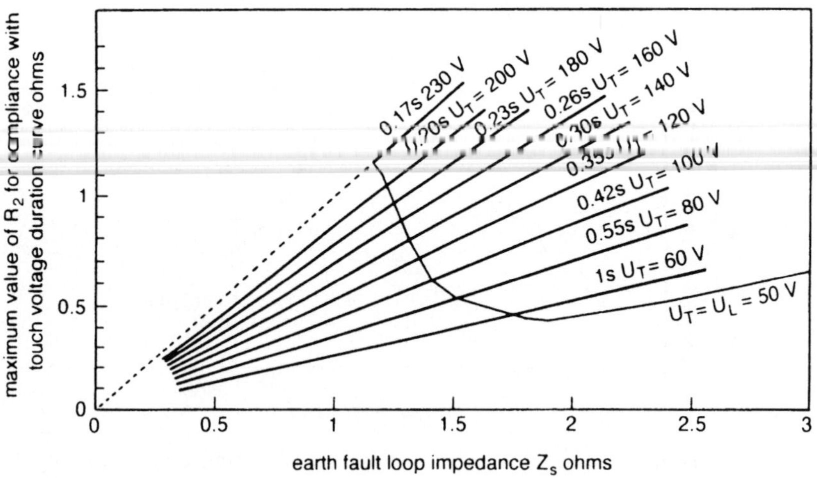

Figure A.12 Impedance characteristic and superimposed touch voltage lines.

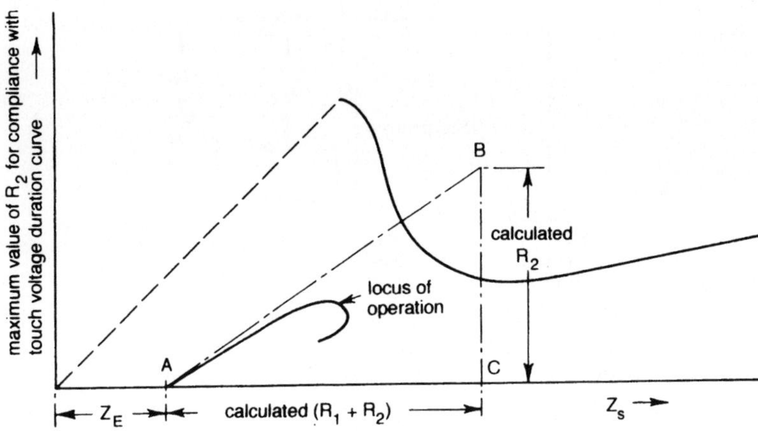

Figure A.13 Showing effect of supplementary bonding.

concept itself) could be used as a design approach and it would be practicable to check compliance by measuring Z_E and $(R_1 + R_2)$.

Certainly for single-phase circuits meeting a 4% limitation in voltage drop there is no difficulty in meeting the touch voltage duration curve and in many cases meeting the R_{2N} limit presents no problem. For three-phase circuits compliance with the touch voltage duration curve is a little more difficult but not impossible.

In practice the problem arises because either deliberate supplementary bonding is used or there is a fortuitous contact between exposed-conductive-parts and extraneous-conductive-parts. For example, returning to Figure A.1, the exposed-conductive-parts of equipment A could be locally bonded to the extraneous-conductive-part, because the metallic enclosure of the equipment could be bolted to a stanchion or other metallic part of the building structure.

In either case the touch voltage between the exposed-conductive-parts of the faulty equipment and such parts of other circuits or extraneous-conductive-parts not associated with the local bonding is far less than the design value and as indicated in Figure A.13, the locus of operation will no longer be a straight line. As shown in that figure the design value is outside the impedance characteristic but, either because of supplementary bonding or a fortuitous earth current path, the circuit actually complies with the touch voltage duration curve. The touch voltage between the exposed-conductive-parts of the faulty equipment and the extraneous-conductive-part to which the bonding has been made will be usually very low.

All the previous comments have related to the shock risk hand-to-hand. Referring back to Figure A.1 the exposed-conductive-parts of the faulty equipment will attain a potential above that of the reference Earth. The floor on which the person is standing is considered to be an extraneous-conductive-part and that person is subjected to the shock risk hand-to-feet when he touches the exposed-conductive-parts of the faulty equipment.

The potential of those exposed-conductive-parts above the reference Earth can be taken to be $I_F(Z_2 + Z_3)$ volts but the current through the person's body hand-to-feet will now be determined by the resistance of the body, that of the floor on which the person is standing, and that of the person's footwear.

Finally, consider the case of final circuits fed from a sub-distribution board, as shown in Figure A.14, where R_{21} is the resistance of the protective conductor of one

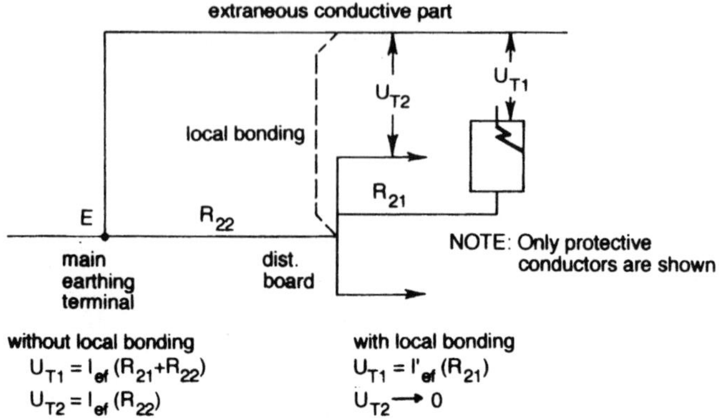

Figure A.14 Touch voltages with and without local equipotential bonding.

of the final circuits and R_{22} is the resistance of the protective conductor to the sub-distribution board.

When there is no local equipotential bonding (or fortuitous path to earth) at the board and the fault has occurred in a final circuit feeding fixed equipment which normally has no limitation as regards touch voltage but should disconnect within 5 s the touch voltage from exposed-conductive-parts of a *healthy* circuit feeding socket-outlets to the extraneous-conductive-part could be such that compliance with Regulation 413–02–04 would not be obtained. Hence the requirements specified in Regulation 413–02–13.

The treatment of touch voltage in TT systems is different to that explained for TN systems but it is sufficient here to indicate that the touch voltage hand-to-hand in the equipotential zone created by the main bonding will be less in the former. But an installation in a TT system may be protected by only one rcd at the origin so account has to be taken of the case where the person protected may be outside the equipotential zone.

Index

adiabatic equation
in earth fault calculations, 60, 93, 94
in short circuit calculations, 94, 119, 120
admixtures of cable sizes in enclosures
calculation of conductor cross-sectional area, 28
already installed circuits
change of parameters, 25
ambient temperature
definition, xiii
symbol for
actual or expected value, xi
correction factor, xi
reference value, xii
armour impedance
calculation of, 88
values of, per metre, 88, 89, 90, 91

bonding conductor
definition, xiii
bunched
definition, xiii

circuit protective conductor
calculation of cross-sectional area, 93
definition, xiii
circuit route length
symbol for, xi
combined examples, 141
conductor cross-sectional area
symbol for, xi
conductor cross-sectional area, calculation for
circuit live conductors, 1
circuit protective conductors, 93
circuits for star-delta starting of motors, 24
circuits in enclosed trenches, 12
circuits in thermally insulating walls, 5
circuits in varying external influences and installation conditions, 6
circuits in ventilated trenches, 9
circuits on perforated metal cable trays, 11
circuits totally surrounded by thermally insulating material, 6
circuits using mineral-insulated cables, 10
motor circuits subjected to frequent stopping and starting, 22
protective conductors, 93
conductor temperature
symbol for
actual operating value, xii
maximum normal operating value, xi
maximum permitted value under overload conditions, xi
conduit impedance
calculation of, 81
values of, per metre, 82, 83
correction factors
for grouped ring circuits, 21
for low ambient temperatures, 20
selection of, 2
current-carrying capacity of a conductor
definition, xiii
symbol for
actual tabulated value, xi
effective value, xi
required tabulated value, xi
cut-off current, 137

design current of a circuit
definition, xiii
symbol for, xi

direct contact
definition, xiii
distribution circuit
calculation of impedance under
earth fault conditions, 77,
short circuit conditions, 140
definition, xiii

earth fault current
definition, xiv
symbol for, xi
earth fault loop impedance
basic equation
for radial circuits, 62
for ring circuits, 62
calculation of
for circuits fed from sub-distribution boards, 77
taking into account transformer impedance, 74
using more accurate approach, 68
using simple approach, 64
when armour is cpc, 88
when conduit is cpc, 81
when trunking is cpc, 81
definition, xiv
symbol for, xii
earthing
definition, xiv
earthing conductor
definition, xiv
earth leakage current
definition, xiv
enclosed trenches, circuits in
calculation of conductor cross-sectional area, 12
equipotential bonding
definition, xiv
exposed conductive part
definition, xiv
external influences
circuits in varying, 6
definition, xiv
extra-low voltage circuits
calculation of voltage drop, 57
extraneous conductive part
definition, xiv

fault current
definition, xv

grouped circuits not liable to simultaneous overload
calculation of conductor cross-sectional area, 13
grouped ring circuits
calculation of conductor cross-sectional area, 21
correction factors, 21
grouping correction factor
definition, xi

impedance characteristics, 169
impedances
symbols for, xii
indirect contact
definition, xv

k values at different initial temperatures, 98

let-through energy
of fuses, 119, 120
of mcbs, 128, 136
live conductors of a circuit
calculation of cross-sectional area, 1
design parameters, 1
live part
definition, xv
low ambient temperatures, circuits in
calculation of conductor cross-sectional area, 19
correction factors, 20

main earthing terminal
connection of conductive parts to, 163
definition, xv
mineral-insulated cables
calculation of conductor cross-sectional area, 10
sheath as cpc, 82
motor circuits
for frequent stopping and starting, 22
for star-delta starting, 24

nominal current of protective device
symbol for, xi

origin of an installation
definition, xv

overcurrent
  definition, xv
overcurrent protective device
  symbol for correction factor for
    type, xi
overload current
  definition, xv

PEN conductor
  definition, xv
perforated metal cable trays, circuits on
  calculation of conductor
    cross-sectional area, 11
protective conductor
  definition, xv
protective conductor cross-sectional area
  calculation of, general, 93
  calculation when protective device is
    a fuse, 96
  calculation when protective device is
    an mcb, 102
  calculation when protective device is
    an rcd, 108

residual current
  definition, xv
  symbol for rated value, xi
residual current device
  definition, xv
residual operating current
  definition, xvi
resistances, of aluminium conductors
  at 20°C, 69
  under earth fault conditions, 64
  under short circuit conditions, 118
resistances, of copper conductors
  at 20°C, 69
  under earth fault conditions, 63
  under short circuit conditions, 118
resistances, of steel wire armouring
  at 20°C, 88, 89, 90, 91
  under earth fault conditions, 90
ring final circuits
  calculation of voltage drop, 55
  correction factor for grouping, 21
  definition, xvi

short circuit conditions, calculations
    related to
  more rigorous approach
    for single-phase circuits, 124
  simple approach
    for single-phase circuits, 117
    for three-phase circuits, 130
short circuit current
  definition, xvi
  symbol for, xi
symbols, xi
system
  definition, xvi

thermally insulating material, circuits
    totally surrounded by
  calculation of conductor
    cross-sectional area, 6
  correction factor, 2
thermally insulating walls, circuits in
  calculation of conductor,
    cross-sectional area, 5
touch voltage concept, 163
trunking impedance
  calculation of, 81
  values of, per metre, 87

ventilated trenches, circuits in
  calculation of conductor
    cross-sectional area, 9
voltage, nominal
  definition, xvi
  symbol for, xii
voltage drop, calculation of
  in ELV circuits, 57
  in ring final circuits, 55
  more accurate approach taking
    account of conductor operating
    temperature, 39
  more accurate approach taking
    account of operating
    temperature and load power
    factor, 53
  more accurate approach taking
    account of load power factor, 51
  the simple approach, 35

COLEG MORGANNWG
ABERDARE
LEARNING
RESOURCES
CENTRE